Organic Dairy Farming

A Resource for Farmers

2006 edition

edited by Jody Padgham

Gays Mills, WI

Copyright © 2006 by Community Conservation

All rights reserved. No part of the contents of this book may be reproduced or transmitted in any form or by any means without the written permission of the publisher.

Library of Congress Control Number: 2005936701

Project Coordination by MOSES
Midwest Organic and Sustainable Education Service
www.mosesorganic.org
715-772-3153, P.O. Box 339, Spring Valley, WI 54767

Our mission is to help agriculture make the transition to a sustainable organic system of farming that is ecologically sound, economically viable, and socially just, through information, education, research, and integrating the broader community into this effort.

Published by Orang-utan Press
The publishing division of Community Conservation
www.communityconservation.org
608-735-4717
50542 One Quiet Lane, Gays Mills, WI 54631

In order to promote global biodiversity and sustainable land use, Community Conservation catalyzes, facilitates and empowers local people to manage and conserve natural resources within the social, cultural and economic context of their communities.

Book Design and Photo Editing by
DragonFire Design
www.dragonfiregraphics.com

The majority of photos in this book have been taken by Harriet Behar, Joyce Ford and Jody Padgham.

The following are used by permission of author:
Feed Quality Preservation, Chapter 2. Originally published in Midwest Bio Ag Newsletter.
A Self Fed Mineral Program, Chapter 3
Holistic Animal Husbandry and Evaluation of Udder Health, Chapter 4
"Pure Delight," Chapter 7
Biodiversity, Chapter 11. Originally published in *Biodiversity Conservation: an Organic Farmer's Guide* by Wild Farm Alliance.

Printed on Recycled Paper

ISBN-13: 978-0-9637982-3-7
ISBN-10: 0-9637982-3-5

Table of Contents

Foreword	IV
Authors	V
1. Introduction (Jody Padgham)	1
2. Dairy Nutrition	8
Understanding Ration Balancing and the Cow (Gary Zimmer)	9
Feed Quality Preservation (William A. Zimmer, DMV)	12
Dairy Nutrition Management (Jeff Mattocks)	15
3. Cow and Calf Management	23
Dry and Lactating Cow Management (Dr. Paul Dettloff)	23
A Self Fed Mineral Program (Dr. Richard Holliday)	27
Managing Dairy Calves (Dan Lieterman)	29
Farmer Profile: Mark Esllinger	39
4. Organic Health Care	41
Holistic Animal Husbandry (Dr. Richard Holliday)	41
Veterinary Tools for the Organic Herd (Dr. Paul Dettloff)	43
Evaluation of Udder Health (Dr. Richard Holliday)	48
5. Stress and Handling	50
Some Thoughts about Stress (Dr. Richard Holliday)	50
Animal Handling (Tamiko Thomas)	52
6. Farm Milk Quality (Louise Hemstead)	59
Farmer Profile: Jim Greenberg	73
7. Farm Biosecurity (Tamiko Thomas)	76
8. Animal Breeding for Organic Production (Robert Hadad)	79
Farmer Profile: Cheyenne Christianson	83
9. A Biological Approach to Soil Health	85
Bio-logical Soil Balancing (Gary Zimmer)	85
Soil Testing and Interpreting Lab Results (Glen Borgerding)	92
Organic Matter Management, Nutrients and Root Health (Walter Goldstein)	96
Farmer Profile: Wayne Peters	100
10. Organic Cropping Systems	102
Understanding Crop Rotations (Mary-Howell Martens)	102
Precision Organic Farming (Mary-Howell Martens)	105
Pest Management in Organic Cropping Systems (Elizabeth Dyck)	109
11. Pasture Management	118
Pasture Management (Laura Paine)	118
Farmer Profile: Francis Thicke	130
Biodiversity (Harriet Behar and Jo Ann Baumgartner)	132
12. Understanding Organic Certification	134
Understanding Organic Certification (Joyce Ford and Lisa McCrory)	134
Record Keeping for Organic Production (Harriet Behar)	147
13. Marketing Organic Dairy Products	151
The Consumer Connection (Jody Padgham)	151
Cooperatives and Marketing Associations (Jody Padgham and Chad Pawlak)	152
On Farm Dairy Processing (Jack Lazor)	155
14. Appendices	i
A. Organic Matter Research (Walter Goldstein)	i
B. Vitamin and Mineral Function and Deficiency in Livestock	iv
C. Alternative Dairy Breeds (Robert Hadad)	vi
D. Livestock Record Keeping Forms (ATTRA)	xi
15. Glossary	xx
16. Resources	xxii
17. Index	xxv

Foreword

At the 2003 Upper Midwest Organic Farming Conference in La Crosse, Wisconsin a small group of folks were standing in the hall chatting (typical behavior for this great networking event), and the conversation turned toward the book *Organic Dairy Farming* by Laura Benson and Robert Zirkel of Kickapoo Organic Resource Network. The supply of this popular book, originally published in 1995 by Community Conservation Consultants (now Community Conservation), had again dwindled to the last few boxes and the issue of reprinting was a topic of concern.

In that hall, Faye Jones, Executive Director of the Midwest Organic and Sustainable Education Service (MOSES), pointed out to Robert Hadad of the Humane Society U.S. and Tim Griffith of Organic Valley/CROPP that the 1995 book was not only still very popular, but was, in fact, one of the only books available to help people get started in organic dairy production. Knowing that the book was long outdated, particularly since the National Organic Program went into effect in 2002, Faye was testing the waters to find support for a total rewrite of the book. The overall conclusion of that conversation was that a rewrite was long overdue. A planning committee was formed of the three organizations to move the project forward.

MOSES stepped forward to act as project coordinator, with MOSES Education Director Jody Padgham serving as book editor. The committee recognized that there was vast expertise on organic dairy production within our community of farmers, veterinarians, researchers and consultants, and so it was decided that the new book would be a compilation of chapters written by these experts. The group concluded that this project would go beyond a simple rewrite of the original publication and vastly expand upon the contents of the first book. We did, however, decide to keep the same name and the same publisher since they had maintained the availability of the book through three printings. Community Conservation, a non-profit conservation organization that encourages communities to protect their natural resources, agreed to work with MOSES to publish the rewritten book.

The committee is very pleased to have had the support and enthusiasm of the twenty authors who contributed their time and expertise to this book. Each is an expert in his or her field. Many have worked very closely with dairy farmers for several years.

We would like to thank the contributors and the many people who read and re-read the text as it was completed. Appreciation to draft reviewers: Craig Adams, Harriet Behar, Roger Blobaum, Joyce Ford, Bridget O'Meara, Joe Pedretti and Jim Riddle.

Many thanks to Joyce Ford for facilitating the indexing process and dedicating many hours to ensuring the Certification chapter represented changes in the law.

Our special thanks go Blooming Prairie Foundation, Organic Valley/CROPP and Organic Farming Research Foundation for their generous financial support which made this project possible.

Organic Dairy Farming Authors

Harriet Behar
Harriet Behar has been an independent organic inspector since 1991. She also teaches organic inspector training courses and workshops for retail food stores about facility certification. An active and organic farmer, Harriet and her husband Aaron have a bedding plant, vegetable and herb farm near Gays Mills, WI.

Glen Borgerding
Glen Borgerding is owner of Ag Resource Consulting Inc., a soil testing laboratory and consulting service based in Albany, MN. Glen is also co-owner of Organic Land Management, Inc., a farm management company based in Chillicothe, MO that specializes in the management of organic farming operations from certification through marketing. Glen has worked with farmers across the Midwest, specializing in nutrient management, organic farming systems and non-chemical weed control since 1984.

Dr. Paul Detloff, DVM
Dr. Paul Detloff is the staff veterinarian for Organic Valley/CROPP. He practiced conventional veterinary medicine for eleven years before starting his own practice in 1982 using biological and organic approaches from the soil up. He is the co-founder of Crystal Creek: A Natural Veterinary Catalog, and founder and owner of Dr Paul's Veterinary Products. Dr. Detloff is the author of four books, including *Alternative Treatments for the Ruminant*. The Detloff's have six children, and live on a tree farm in rural Arcadia, Wisconsin.

Elisabeth Dyck
Elizabeth Dyck conducted research on certified organic acreage at the University of Minnesota's Elwell Agroecology Farm and with farmer cooperators in southern and western Minnesota. Before working in Minnesota, Elizabeth worked with small-holder farmers in Kenya to restore and sustain soil productivity. She is currently working on a manual for farmers wanting to convert to organic production and is living on her newly purchased farm in Bainbridge, New York.

Joyce E. Ford
Joyce E. Ford has been inspecting organic farms and livestock operations for 12 years. She is a Training Coordinator for the Independent Organic Inspectors Association and has co-authored several inspection manuals and other publications. She currently serves on the MOSES Board of Directors.

Walter Goldstein
Walter Goldstein is Research Director at Michael Fields Agricultural Institute, and also manages a small sheep farm with his family. He and his colleagues

are engaged in research at the Wisconsin Integrated Cropping Systems Trials and in on-farm research in Wisconsin and Illinois. Goldstein also has active research interests in biodynamic preparations and in open pollinated corn.

Robert Hadad
Robert Hadad is currently the Director of Farming Systems-Farm Animals and Sustainable Agriculture for The Humane Society of the United States where he conducts education and outreach to family farmers and consumers to: reexamine food production systems, improve quality and profitability of family farms, improve animal welfare through humane sustainable agricultural practices, improve marketing and create greater localized support for family farms in this country. He has previously been a research faculty member with the University of Minnesota investigating sustainable organic farming system transition strategies and alternative crop marketing and one of the first sustainable agricultural Cooperative Extension Educators working in KY and NC.

Louise Hemstead
Louise Hemstead is the Chief Operating Officer at CROPP Cooperative, a farmer owned national cooperative headquartered in LaFarge, WI. She is responsible for farm milk pickup to customer delivery including relationships with co-processors who convert organic farm milk into organic dairy products. She and her husband operate an organic dairy farm which has been in her husband's family since 1893.

Dr. Richard Holliday, DVM
Dr. Richard Holliday received his DVM degree from the University of Missouri in 1959. For the last 18 years he has been employed as Technical Services Veterinarian by IMPRO Products, Inc., a company that produces and markets holistic animal health products for dairy cattle. Dr. Holliday became certified as a Veterinary Acupuncturist in 1988, and has been actively involved in promoting organic agriculture and holistic veterinary medicine for over 30 years.

Mary-Howell Martens
Mary and her husband Klaas Martens farm over 1300 acres organically in upstate New York. The Martens are well known for their success as large-scale diversified grain farmers with an emphasis on soil fertility and innovative weed control. Mary has been a frequent contributor to the NewFarm website and is a regular speaker at conferences throughout the East Coast.

Jack Lazor
Jack Lazor owns and runs Butterworks Farm in North East Vermont with his wife Anne. Farming since 1979, in 1984 the Lazors became certified by the Vermont Department of Agriculture for on-farm processing of milk and yogurt. Butterworks Farm now distributes products from their 43 cow Jersey herd throughout the East Coast. Jack is a frequent speaker at conferences on topics relating to on-farm processing and organic dairy management. The Lazors have been important mentors to numerous farm scale dairies.

Jeff Mattocks
Jeff Mattocks is a nutritional consultant for The Fertrell Company in Bainbridge, PA. Jeff has been working with natural and organic farmers for the past 10 years, developing and balancing diets. His primary focus is organic dairy and poultry production.

Lisa McCrory
Lisa McCrory is the Dairy and Livestock Advisor for the Northeast Organic Farming Association of Vermont.

Jody Padgham
Jody Padgham has been the Education Director at the Midwest Organic and Sustainable Education Service (MOSES) since 2001. Previous to her current position she worked with farmers as an Outreach Specialist for the University of Wisconsin Center for Cooperatives. Jody owns a 60-acre grass farm in North Central WI where she raises organic chickens, turkeys, laying hens and sheep.

Laura Paine
Laura Paine is an agronomist and an Agriculture Agent with the University of Wisconsin Extension Service, based in Columbia County. She has a broad plant science background with education and training in botany, horticulture, and agronomy. For the last 12 years, she has been involved with research and education in the area of grazing management including resource conservation issues such as water quality, wildlife habitat, and using native prairie plants in pasture systems.

Chad Pawlak
Chad has been involved in value added agriculture since 2001. Currently Chad serves as President of Wisconsin Organics, Inc.. Chad has assisted over twenty-five dairy producers in organizing themselves into a competitive organic milk pool. Chad has a strong commitment to the family farm and working to expand the Wisconsin Organics (www.wiorganics.com) brand throughout the Midwest market they service.

Tamiko Thomas
Tamiko Thomas works in the Farm Animals and Sustainable Agriculture section of The Humane Society of the United States. Previously she lived and worked for a number of years on a small family dairy and at the University of British Columbia's Dairy Education and Research Centre as a research assistant. She has a BSc in Agriculture (Animal Science major) from the University of Alberta in Canada and a MSc in Applied Animal Behavior & Animal Welfare from the University of Edinburgh in Scotland. Her dissertation was on the vocal behavior of dairy calves.

Gary Zimmer
Gary Zimmer is president of Midwestern Bio-Ag, author of the recent book *The Biological Farmer*, and the owner, with his family, of Otter Creek Organic Farms in SW Wisconsin.

Chapter 1

Introduction

Why would I ever go back to conventional dairy farming? Even if I lose the organic premium, I'll never stop farming organically. My cows are healthier and more productive, the crops yield higher quality feed, I have hardly any weed pressure anymore – things are only continuing to get better." – Jerry Wagner, ten-year Certified Organic Dairy Farmer, Black River Falls, WI.

If you are dairy farming now or are interested in dairy farming in one of the dairy states in the U.S., chances are that you know of someone who is already an organic producer, or considering becoming one. With surging markets and strong profitability, organic production is a viable option for many dairy farmers. Those wishing to run small herds or stay on family farms are finding organic production to be an especially good choice. With the consumption of organic dairy products steadily growing at over 25 percent per year, the industry is stable and in need of new producers. Making the change to organic production will take time and commitment; but it can be a straightforward and highly rewarding process.

What is organic dairy production?

Since first domesticated 8,500 years ago, cows have been used to produce milk for human use. Modern dairy cattle, such as those used in the U.S., descended from northern European native breeds in the late 1800's. Improvements in dairy husbandry, crop production systems and milking systems have occurred as populations and demand for milk, cheese and other milk products grew. For the first 8,450 years of dairy farming, dairy products were successfully produced without the use of petroleum-based chemicals.

The 1940's and 1950's were pivotal times for agriculture in the U.S. Excess chemicals developed for WWII were brought home and used to create "miracles" on farm fields.

Chapter 1 Introduction

Modern organic production grew out of a reaction to the "more is better" mentality that was casually, and often innocently, adopted by farmers in regard to pesticides, fertilizers and herbicides. Adverse reactions to chemical use, including illness, cancers and environmental degradation, created a movement of people desiring to go back to growing food without chemical inputs, using natural systems as support. In the 50 years since, much has been learned about creating balanced organic systems. The same set of beliefs and values that got the organic movement started still governs our choices today.

From 1960 through the early 1990's the definition of "organic production" began to coalesce. Throughout this period, individual states, countries and independent certification agencies set standards for producers to meet in order to qualify for the use of a number of independent "certified organic" labels.

In October 2002, the U.S. Department of Agriculture took over the regulation of organic labeling in the U.S. The Federal Organic Foods Production Act, passed in 1990, was implemented in 2002 after years of hard work to come to agreement about a national standard that would define exactly what farmers must do to call their product "organic." Farmers, researchers, consumers and government representatives worked together to write the rules of this complicated and comprehensive law.

Organic farming is all about producing quality food while protecting resources and maintaining optimum animal health.

The law, the Organic Foods Production Act (OFPA), created the National Organic Program (NOP). The NOP is detailed and serves as the "bible" for organic farmers. The NOP requires that a USDA accredited certifying agent inspect and review the records and approve the plans and methods of any farm wishing to label products as organic.

According to the NOP, the definition of "Organic Production" is: "A production system that is managed to respond to site-specific conditions by integrating cultural, biological and mechanical practices that foster cycling of resources, promote ecological balance and conserve biodiversity." We will describe in this book the details of creating sustainable organic dairy production systems that uphold this definition.

Organic dairy in the U.S.

Increasing consumer demand is driving a steady growth in the organic market, with dairy the fastest growing products in organic food sales. The strong growth in the organic dairy market has fueled a sustained interest in organic dairy production, including the transition to organic of numerous conventional systems each year. Demand is expected to continue to grow, as consumers maintain a strong interest in organic products for health and environmental reasons. New certified organic dairy farms around the U.S. are needed to supply these growing markets.

Organic milk first appeared in supermarkets in 1993, and now organic milk, cream, half and half and cheese are available in many grocery stores. Sales of organic dairy in mainstream grocery stores are growing 36 percent per year. In 2000, $480 million of organic milk, cheese, butter, eggs and yogurt were sold in the U.S., with organic milk and cream being the second leading category of organic food sales in conventional markets. [1]

The Economic Research Service also reports that in 2001 there were 2,341,482 certified organic acres in the US, showing a growth of 74 percent since 1992. In 2001 there were 48,677 certified dairy cows in the U.S., with 10,803 of those living in Wisconsin alone. California showed 9,251 certified organic cows and New York 6,704. The number of certified organic milk cows increased 27 percent between 2000 and 2001, and as of 2001 there were 492 certified organic dairy farms in the U.S., which is 1.2 percent of the total dairy farms in the country.

Organic dairy products have become common on grocery shelves.

Most of the organic milk produced in the U.S. is marketed through a few companies (including the cooperative, Organic Valley/CROPP, headquartered in Wisconsin, and the privately owned company, Horizon Organic Dairy, of Colorado.) Several other smaller regional cooperatives and marketing pools have developed in the past few years. Some producers process organic products on-farm and market directly to consumers.

Why do farmers choose organic farming?

In the early years of organics many farmers carried very strong ethics with them into the barnyard as a motivation for organic production. Resistance to chemical use, a desire to see improved herd health and interest in fulfilling a growing consumer demand fueled organic transitions. With improved market access and stable high prices, organic production has become desirable to a larger population whose primary impetus may be to achieve higher profitability, often while maintaining a moderate herd size. Many farmers have found that a transition to organic dairy production has allowed them to keep the family farm profitable. Strong milk premiums and expanding markets support a larger milk check, especially when combined with the lower input costs of rotational grazing.

---

[1] *USDA-Economic Research Service, "Recent Growth Patterns in the U.S. Organic Foods Market" Agricultural Information Bulletin Number 777. September 2002*

Chapter 1 Introduction

Organic products generally receive a premium in the marketplace. For organic dairy farmers, that currently translates to approximately $18-$21 base price per hundredweight for fluid milk. A number of factors are computed in the price of organic milk, including butterfat content, protein, other solids, and quality, which can add a premium of up to five dollars per hundredweight. An economic comparison of organic and conventional dairy systems found that the average organic dairy operation netted $477 per cow vs. $255 per cow for the conventional dairy operation of similar size. [2] Although organic dairy farms had 21 percent higher expenses overall (mostly feed and feed supplements), veterinary costs, labor, freight and trucking were significantly less. Do the math for your operation!

How to get started

Organic dairy farming is based on a foundation of soil management, quality crop and forage production and low input animal husbandry. Many agree that becoming a successful organic farmer means that you need to start thinking about your farm — the land, the crops, the animals — in a new way. Solutions to problems are planned with an emphasis on prevention, rather than as remedies administered "after the fact." Every decision made on the farm, regarding soil management, crop rotations, pasture use … is planned toward the end of improving the health of the milking cow herd. A whole systems approach is the key to successful organic farming.

> A great resource that helps producers "do the math" is *The Organic Decision: Transitioning to Organic Dairy Production Workbook*, published by Cornell University. Check out www.cornell.edu or call 607/254-7412 to purchase a copy.

Accurate record keeping is also essential to the successful organic farm. Functional recording methods must be established and followed to retain organic certification. Comprehensive records have multiple benefits to the organic producer and can be invaluable to guide production choices and long-term management decisions.

One organic farmer recently said that he liked organic production because "it made farming fun again." "You get to try things, learn and make decisions again," he said. "You don't just follow what the chemical guy says."

If "fun farming" in a dynamic environment of learning and change sounds interesting, then this book is for you. We have brought together twenty experts in the field, including veterinarians, farmers, researchers and specialists, to share their expertise. We will take you through the general concepts and many details of developing a successful organic dairy system.

---

[2] McCrory, Lisa, "An Economic Comparison of Organic and Conventional Dairy Production, and Estimations on the Cost of Transitioning to Organic Production", Northeast Organic Farming Association of Vermont, May, 2001, page 2. (This report is based on data collected in 1999. The program used for this study did not allow separation of purchased hay from purchased grain and feed supplements. They were not able to put a value on average age of the herd or how much of cattle sales was livestock replacement, calves or culls.)

Organic production will not be for everyone. But as Jerry Wagner says, those who have tried it and liked it will never "go back." We invite you to dig into the following pages, explore organic production and decide for yourself whether it might work for your farm and family.

As you read the book you may notice that some of the topics presented have been approached from different directions by different authors. For instance, in Chapter Three Dan Lieterman carefully describes ideal calf care conditions, which involve the standard practice of separating calf and mother soon after birth. But, in the Farmer Profile of Mark Eslinger you will read that Mark leaves his dairy calves with their mothers for up to three months. Mark has found a unique management that works for his particular situation. The authors of this book have written what they recommend and have learned from experience. Although there are certainly standard practices and procedures, there is no one "right way" to farm. Successful farming is about adaptation and experimentation. We offer you the ideas in this book with the hope that you will expand, modify and refine them to fit your own particular situation.

It is a temptation to begin this book by orienting you to organic production "from the ground up," since that is the way we want you to start thinking about your farm. But knowing that the typical dairy person will be wondering first about the changes they need to make in the barn, we'll start by looking at the cows.

This book begins with a series of chapters that focus on cow management. We start with a look at dairy nutrition, and move from there to cow and calf management and the basics of organic veterinary health care. Then we'll talk about stress management, maintaining milk quality, biosecurity and breed selection.

Next we will help develop your understanding of organic soil management, because the soil and what you grow on it is the base of your success as an organic dairy farmer. The final sections of the book explore building soil health, organic cropping systems and pasture management. We conclude with a full explanation of organic certification, and a section on marketing organic dairy products. Several appendices support the text, including detail on organic matter research, vitamin and mineral function and deficiency tables, a section on alternative dairy breeds, examples of on-farm records for certification, a glossary, useful resources and index.

There can be no better learning environment than a sustainable organic farm. If you decide to take the next step to organic dairy production, we encourage you to attend farm field days and do your own visits to organic dairy operations in your area. Consider an internship or mentor relationship with an experienced organic farmer. There is no teacher like experience. See the resources section for suggestions on how to find field days, pasture walks and other farmers in your area.

Understanding Dairy Cows
Jody Padgham

Ask dairy people what the key component of their operations are and the easy answer will be the dairy animal. The cow barn is the heart and soul of a dairy business. Keeping the cows healthy and productive is obviously the cornerstone of a good dairy farm.

Those with organic operations give a lot of attention to the cow barn, but do not stop there. Organic farmers spend as much time managing soil nutrition and crop health as cow health. Organic farmers maintain cow health by providing the base proteins, fibers, minerals and vitamins that a cow needs to stay healthy and productive. Successful organic dairy farmers must see their cows as part of a living, holistic farm system rather than as milk producing machines.

Many who look at organic dairy production will have among their first questions, "What can I use instead of (conventional product of choice) to treat my cows for (common disease or problem)?" Long time organic inspector and trainer Harriet Behar from Gays Mills, Wisconsin answers that question with this statement: "Organic isn't about substituting approved products for chemical products. Organic is about managing your system for maximum soil, plant and animal health so that you don't need to use very many 'products' or off-farm inputs."

Any one who has been in dairy for a while knows what it takes to keep a cow healthy, right? Well, maybe not. When is the last time your vet came in and said something like "What do you think is going on in the herd to cause this large number of DA's (displaced abomasums)?" It is not uncommon for traditional dairies with 80 to 100 cows to have one or more DA's per month. Expensive and stressful, generally requiring surgery, DA's are considered fairly standard fare in the conventional dairy, and part of the process of managing cows for milk production. However, the average organic farmer rarely sees a DA. What does this tell us? There is something occurring within a conventional system that is contributing to the occurrence of DA's in the herd. And something within an organic system helps cows to remain healthy enough to avoid DAs. So again, back to the basic question: What is it we need to do to maintain a cows optimum health?

Basic herd health

There are some basic elements in keeping cows healthy.
1. Quality nutrition — access to feed and/or supplements containing appropriate carbohydrates, proteins, minerals and vitamins. Energy expended in getting nutrients must be overbalanced by the nutrients received in order for the cow to thrive, develop calves and give milk. The organic standards require that all ruminants have access to pasture that provides significant feed value.

2. Access to clean water
3. Maintenance of a strong immune system and treatment for any extreme health situations or issues.
4. A low stress environment.
5. Genetics appropriate to the situation.

The chart in box A offers a visual picture of how these elements fit together. The "Epitome of Health" will equal the "Apex of Production" when all of the listed elements of a farm system are balanced. In the following chapters we will help you understand how to optimize these elements for the organic cow.

We will not go into detail on water issues. Access to clean, untainted water is essential to maintain animal health. Regular water tests should be taken to prove that water quality is pure. When careful manure management, erosion prevention and soil health practices are followed (all required under organic production) basic water quality is generally maintained.

Healthy Herds   Dr. Paul Dettloff

To maintain a healthy animal in the organic arena producers must step back and look at their entire farm system. The system can be compared to a pyramid that includes the entire farm. The apex of health, or the epitome of health, is where a healthy immune system meets with good stable production.

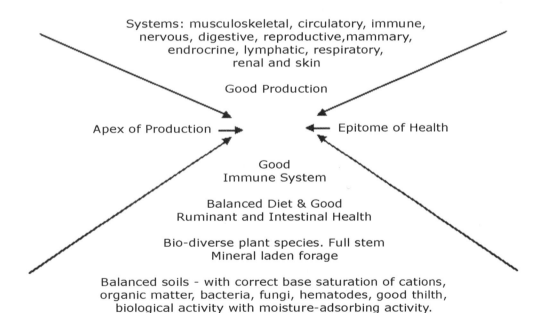

By taking the holistic approach and using the entire set of tools listed in the lower triangle of our pyramid, we strive to build up all of the systems listed in the upper triangle.

Chapter 2

Dairy Nutrition

Successful dairy production is based on optimizing the health of the dairy animal. In her lifetime a cow will cycle through several life stages. From her beginnings as a calf to maturation, breeding, pregnancy, freshening, milk production and then her dry stage many things change but several stay constant.

As one of the largest ruminants, a cow's digestive system is unique. While most animals only have one stomach, ruminants have four stomach compartments (the rumen, the reticulum, the omasum, and the abomasum, or true stomach.) The ruminant system has two significant advantages over a single stomach: more space for processing large quantities of forages, and an environment that is ideal for enormous populations of microorganisms. A typical cow rumen will hold twenty five to eighty billion bacteria per ml, and 200,000-500,000 protozoa. The number of bacteria will vary depending on the nature of the diet, the feeding regimen, species differences, individual animal differences, the season and the availability of green feed.

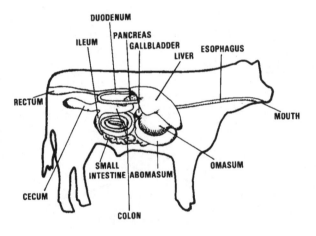

Cow digestion

Another difference between a ruminant and a non-ruminant is the actual act of rumination. We all are familiar with the calm, slow, blissful looking cow lying under a tree chewing her cud after a hard day of grazing. During rumination the cow actually regurgitates and rechews a soft chunk of coarse feed (called the bolus). Each bolus will be chewed for about a minute and then re-swallowed. A ruminant will spend as much as 8 hours per day chewing her cud. A coarse, fibrous diet will lead to longer ruminating. The lower the fiber content, the more digestible the feed, and the less time needed to break down the feed. Thus, a higher quality forage (less fiber) means a cow can eat more in a day. Rumination also stimulates saliva production in the mouth, which helps food to pass through the digestive tract. In addition, saliva buffers the pH of the rumen, which protects against acidosis. Through the active microbial fermentation in the rumen, ruminants produce and pass more gas (mostly methane and CO_2) than non ruminants.

Let's take a little closer look at how the digestive system of a cow functions, and how we make decisions about what to feed that cow to optimize her health.

Understanding Ration Balancing and the Cow
Gary Zimmer

How is the cow designed?

The rumen, the cow's fermentation vat, can hold over 30 gallons of material containing 10-20 percent dry matter. The rumen serves as a reservoir of feed and supports an active microbial fermentation. Bacteria in the rumen attach to feed particles, breaking them down and producing volatile fatty acids as an energy source for the dairy cows. On the average, carbohydrates contribute 70-80 percent of the diet dry matter while protein, fat and minerals make up the remaining portion.

Carbohydrates are the primary energy source for rumen microbes. Two categories occur in feeds: nonstructural and structural carbohydrates. Nonstructural carbohydrates are sugars, found in the cells of growing plants, and starch, which is the form of energy found in cereal grains. Structural carbohydrates are cellulose, hemicellulose and pectins, which provide rigidity and strength to plants. They are found in plant stems and structure.

Weather, genetics, maturity and fertilizer inputs all influence the type and amounts of carbohydrates in plants. Some carbohydrates are more digestible than others. As plants get more mature, the non-carbohydrate polymer lignin forms, which is not a digestible material for the rumen.

How does the rumen function?

The rumen of healthy cows will have two contractions per minute. This mixes rumen contents, brings bacteria and feed into contact, moves material out of the rumen if it is dense and short, and brings long particles to the top surface for rumination to occur. The cow should spend 8-10 hours a day ruminating (cud chewing). The cow re-chews, adds up to 20 gallons of saliva each day (this is her buffer), then re-swallows. With proper forage levels and particle length, supplemental digestive aids are not needed.

The cow mixes the "vat", controls the temperature, breaks down particle length and adds saliva to control pH. You, the dairyman, have control of what goes into the "vat".

What should go into the "vat"?

Holstein cows with good production can, on an average, eat approximately 50 pounds of dry matter feed per day. All cows are different. One advantage of a Total Mixed Ration (TMR) is that it can be balanced for efficient rumen function even with individual cow feed intake variations. It also provides feed uniformity.

A lot of research has been done to improve production by adding things to the fermentation "vat pile." Very little has been done on changing the composition of the "pile" by starting with the soil, a key management tool in organic production. Once the nutritional value of pile ingredients changes, either through composition or quality, all the numbers used to determine ration balance change. Given the variations in soils and thus feed quality, it is hard to come up with the "perfect ration." Testing, as well as calculating and following guide-lines, can be helpful, but these only serve as guides. Observation of animal health and vigor is the best way to assess the effectiveness of your overall ration balance.

What are the essential pieces in the "pile"?

Some feeds can be dry, while others have moisture in them. Converting feeds to a dry matter basis simplifies ration balancing. Pounds as fed times the percent dry matter equals the dry matter pounds intake. For example:

If a cow is fed:	Pounds Dry Matter (DM)
5# dry hay @85% dry	4.25
50# haylage @ 50% dry	25.00
18# corn silage @ 33% dry	6.00
15# dry corn @ 87% dry	13.00
2# roasted soybeans @ 88% dry	1.75
Total	50.00 lbs

Forage

Forage is defined as vegetable matter in a fresh, dried or ensiled state (ie pasture, hay, silage). In the dry state, forages average more than 18 percent fiber. The term "roughage" is often used interchangeably with forage, although roughage usually implies a coarser, bulkier feed than forage. High forage ration feeding starts with a minimum of 60 percent forage, with some farmers feeding up to 75 percent. What is the percent forage in the above ration? Hay is forage, haylage is forage, corn silage is half forage on a dry basis, and the rest of this ration is concentrates. So…

DM hay	4.25
DM haylage	25.00
DM corn silage	3.00
Total	32.25 # forage dry matter

For a 60 percent forage diet you take the Holstein's 50 pound DM intake times 60 percent, which means that 30 pounds of the "pile" needs to be forage. In the example we have 32.25 pounds of forage dry matter in our 50 pounds, or 32.25/50 lb = 64.5% forage. This puts us within a reasonable range in this example.

What percent protein?

There is some protein coming from each feed ingredient. The ration needs to be 16-18 percent protein. 50 pounds DM times 16 percent equals 8 pounds of protein per day.

In the example:
Hay at 18% protein DM basis is	4.25# x 18% = 0.77 lbs.
Haylage at 20% protein DM basis is	25.00# x 20% = 5.00
Corn silage at 8% protein DM is	6.00# x 8% = 0.48
Corn at 10% protein DM basis is	13.00# x 10% = 1.30
Roasted beans at 40% protein DM is	1.75# x 40% = 0.70
Total	8.25 lbs.

So, 8.25% protein/50# feed equals 16.5% protein diet. This is within our range.

How do you calculate energy values?

Energy is not really measured in feed testing. Rough fiber levels from ADF (Acid Detergent Fiber) and NDF (Neutral Detergent Fiber) extraction are measured, and all other numbers such as TDN (Total Digestible Nutrients) and NEL (Net Energy for Lactation) are calculated.

Energy from starch is easily measured, as it is just the amount of grain. Minimize starch--don't overdo it! Starch feeding reduces rumen efficiency, which means there's less being digested from the pile.

ADF, NDF, TDN, and NEL can all be easily calculated. For these calculations, pounds of DM fed are multiplied times the percent of calculated fiber divided by the energy value of each feed. Make a chart for yourself like the ones above, stay within recommended guidelines and you'll be fine. (To find energy values and forage percent, check an animal nutrition textbook in your local library.)

Grain ration supplements forage for heifers.

Ration Energy Target Levels
ADF   >19%
NDF   >28%
TDN   70-74%
NEL   0.74-0.78

What if you only have low quality forages?

If dairy roughages are not of high quality, it is very difficult to supplement your way to profit and success, since they make up over 50 percent of what a cow eats every day.

One option used to make up for poor forages is to feed a higher percentage of concentrates and less roughage. This may yield more milk, but will be at great risk to your cows' health and long-term production. Some people buy in high quality forage, but this will be expensive in the long term and will not lead to a profitable success at dairying. The best solution I recommend is this: Learn to grow high quality forages!

In summary, use common sense in feeding. Be consistent, allow time for changes, and make changes slowly. You have to earn the right to have healthy cows, with high production and very little ration supplementation. Become a fanatic about producing quality forages. Recognize the quality from what you have done to the land and in harvesting, so when you put feeds in your storage system you know what results to expect from feeding them. Take responsibility for your own farm's success- that is management and the fun of farming.

Feed Quality Preservation
William A. Zimmer D.M.V

Preservation of quality requires feed to be stored in specific conditions under which spoilage organisms cannot work to deteriorate the feed quality. Feed can be dried to below 10 to 15 percent moisture, where spoilage organisms are not active. Higher moisture feeds such as high moisture grains (20 to 30 percent moisture) or silages (50 to 70 percent moisture) require the presence of substances that inhibit spoilage organisms.

Conditions that benefit proper preservation of high moisture feeds include eliminating air and oxygen, proper moisture, proper supply of plant sugars and carbohydrates for ensiling organisms, and minimizing the number of spoilage organisms present. Most of these conditions can be achieved by harvesting proper maturity feeds, timely wilting and field drying, properly removing feeds from the field and placing them into storage structures to minimize contamination with soil particles, properly chopping large feed particles to

If silage inoculants are used, be sure they are approved for organic production.

allow maximum packing of feed in the structure and proper harvest equipment maintenance to condition feed. Many storage structures are suitable if they are managed properly, including upright silos, bags, wraps, bunkers, pits and trenches. Piles, trenches and pits are hard to manage. Bunker silos are often mismanaged. All of these structures should be tightly covered to minimize surface spoilage.

There are basically two types of products commercially available for the purpose of improving high moisture animal feed preservation, preservatives and inoculants. Silage preservatives are concentrated sources of organic acids such as acetic, formic, lactic and/or propionic acid. Most commercial preservatives are synthetic and therefore not acceptable for organic production systems. Many inoculants, however, are suitable for organic systems.

Silage inoculants contain live, viable, naturally occurring microorganisms. Selected organisms produce large amounts of organic acids when added to feed at the start of the ensiling process. Acid type and levels produced depend on the type of organism and the conditions of the silo.

Common beneficial organisms found in inoculants include lactic acid producing organisms (e.g. *Lactobacillus, Pediococcus*, etc.). Some inoculants contain organisms that produce propionic acid (e.g. *Propionibacterium*). Most research indicates that a minimum of 100,000 colony-forming units of lactic acid bacteria should be supplied by an inoculant. Specialized inoculants should supply about 500,000 CFUs of propionic acid producing bacteria. Additional active ingredients in inoculants include bacterial or fungal enzyme sources that will begin the breakdown of plant carbohydrates to feed the inoculant organisms. This may result in a faster or more complete fermentation under some conditions. A quality inoculant can reduce storage losses dramatically, especially when ideal conditions for proper fermentation have not been met.

Lactic acid producing organisms are split into two groups, homofermentative and heterofermentative. Homofermentative bacteria primarily produce lactic acid. This acid represents the most efficient conversion of plant carbohydrates to acid. Heterofermentative organisms produce nearly equal levels of lactic and acetic acid. When acetic acid is formed, energy is lost from the fermentation as carbon dioxide. On the other hand, once feed is removed from storage, acetic acid tends to maintain feed stability longer than lactic acid does. Bacteria naturally found on plants are predominantly heterofermentative. Most silage inoculants include predominantly homofermentative bacteria. Whichever strains dominate the fermentation will determine whether feed is more efficiently fermented or is stable longer once removed from storage. Generally, feed that is fed in a timely manner from a properly sized and managed storage structure will benefit from a homofermentative inoculant. Feed that is to be left in the feed bunk for extended periods of time or when the face of the silo can not be properly managed may benefit from an inoculant that enhances heterofermentation.

Propionic acid is better than both acetic and lactic acid at extending feed stability after removal from storage. However, in order for propionic acid to be formed during storage, a propionic acid producing organism must first convert lactic acid into propionic acid. This requires a two-step fermentation and much longer period of time. Propionic acid can not be used in organic production, as it is prohibited by the NOP.

Properly stored feed that has undergone a lactic acid fermentation should reach a pH of less than 5 (<4 for corn silage), contain about 2 to 6 percent lactic acid and 1 to 4 percent acetic acid, with lactic acid making up at least 50 percent of the total organic acids. Butyric acid should be low, as this acid decreases feed palatability. Feeds undergoing a secondary propionic acid fermentation should reach one percent propionic acid levels for best stability. Ideal moisture levels are 65 to 70 percent for proper storage fermentation.

Inoculants should be used to maximize fermentation efficiency and stability. They should not be expected to overcome poor management, forage that has been left in the field too long or poor storage structures. Inoculants help to better maintain whatever quality level is present, but should not be expected to improve forage quality over that present at initial cutting.

Dairy Nutrition Management
Jeff Mattocks

There is no one set of requirements that meets the nutritional needs of cows at all stages of lactation/dry period. In each life stage of a dairy cow, all of the nutrient requirements change: each period of lactation and gestation have critical issues that will affect the next stage. The following sections will outline the basic nutritional requirements for dairy cattle during the various periods of lactation and gestation.

The following ration templates are not intended to replace properly balanced rations and are merely guidelines for educational purposes. These templates are not perfectly balanced for all circumstances and should not be used without adjustment to individual farm circumstances. Requirements outside of the information covered here should be calculated and reviewed by an experienced nutrition advisor.

Feed rations are extremely variable based on the quality of the forages being fed. The importance of high quality forage cannot be emphasized enough. The higher the quality of forage, the less grain required to balance the diet. The younger a forage is cut, the more nutritional value will be preserved in the stored feed, meaning higher protein, energy, and mineral content. All of these factors will contribute to better animal health, productivity and fewer off-farm purchases. You should note that animal health is listed here before productivity. A healthy animal will always provide the highest productivity possible without being pushed. A push to higher productivity does not support a healthy animal. You should always remember that the goal of good health should be first priority and production second priority. With this in mind you will maintain the highest level of profitability. See chart listing common illnesses with corresponding cost associated with each.[3]

Health Problem	Cost	Average Incident Rate	Realistic Goals
Milk Fever	$334	5.30%	< 3%
Ketosis	$145	4.70%	3-5%
Retained Placenta	$285	9.00%	3-5%
Displace Abomasums	$340	4.50%	< 3%

[3] Note: *This information is within 20% accuracy. This information was provided by: The Dairy Focus, Volume 3, Issue 2, February 2002 Penn State, Cooperative Extension. Cost Values Researched by: Dr. Chuck Guard, Cornell University. Average Incidence Rate Researched by: W.S. Burhans, Cornell University, 1999, reported in Total Dairy Nutrition. Realistic Goals Established by: W.S. Burhans, Cornell University and a phone conversation with Dr. Robert Van Saun, Penn State Extension Veterinarian.*

Useful ratios

The use of ratios helps in understanding the basics of a cow's nutritional needs. The most important nutritional requirements to track are energy to protein ratios, forage to grain ratios and calcium to phosphorus ratios.

Energy to Protein Ratio
Without sufficient protein, energy inputs may not be properly utilized and may flush through the system undigested or be converted to stored fat. The alternate imbalance (excessive protein) will cause diarrhea. Diarrhea causes essential nutrients to flush through the digestive tract without being properly absorbed or utilized. Excessive protein does more than just cause stool looseness. If allowed to continue for an extended period of time, it causes kidney problems and the mobilization of stored fat reserves at an accelerated rate.

Forage to Grain Ratio
It has been proven over time that the higher the forage base in a balanced diet the better herd health and productivity will be. Therefore, I recommend that you never go under 60:40 forage-to-grain ratio. My experience is that ratios with higher grain may shorten the life expectancy of the animal through metabolic disorders previously mentioned. The optimum ratio appears to be 70:30 forage to grain, when forages are of a high enough quality to support this.

Calcium to Phosphorus Ratio
These values will dictate the animal's reproduction, milk production and longevity. They will also directly affect the proper absorption of other minerals in the cow's diet. The values should change with different stages of production to insure that the cow's needs are met based on the body functions being performed.

Nutritional requirements for different production stages

Freshening Stage (0-30 days)
During this period the cow will be under the most stress of her entire lactation. During Postpartum she is trying to make milk to feed her calf as well as heal her reproductive tract after calving. The first 30 days after freshening will set the foundation for the remainder of the lactation regarding both health and production. She should have ample amounts of dry, long-stem hay to cud and manufacture buffering agent via saliva to keep the digestive tract stable. Also during this period it is important to maintain a higher forage-to-grain ratio to avoid stomach upsets. The sample ration will provide basic guidelines for nutritional requirements during the "Just Fresh" or Postpartum stage of lactation. The basis for the ration is a ten-day fresh cow, at 40 pounds of 4.0% fat, 3.2% protein of milk production and a Body Condition Score[4] of 3.5.

---

[4] *Body Condition Scoring can be an effective management tool for evaluating the energy reserves of cows and the overall nutritional program throughout the year. UC Davis Veterinary Medicine Extension has a useful website on Body Condition Scoring in dairy cattle: www.vetmed.ucdavis.edu/vetext/INF-DA/INF-DA_BCS.html*

Ration Name: Freshening Ration
Cow Profile Lactating Lactation # 2

Feed Name	As Fed	Dry Matter
Mixed Grass Pasture	70	18.9
Mixed hay	6	5.22
Organic Corn grain ground	10	8.8
Organic Molasses black strap	0.5	0.375
Limestone	0.2	0.198
Ca 23%:P 18%	0.063	0.061
Vit ADE Example	0.095	0.093
Salt-white	0.2	0.198
Mag oxide	0.063	0.062
Selenium 0.06%	0.02	0.02
Concentrations		
Totals	87.141	33.927
Requirements	0	33.549
Difference	0	0.38
Forage NDF as % of NDF	93.9	
Forage NDF as % of DM	38.33	
NDF intake as % of BW	1	
Total forage ADF (lbs)	7.5	
2. Forage:Concentrate ratio	[71:29]	
3. Calcium:Phosphorus ratio	1.9:1	
Potassium:Calcium ratio	2.5:1	
Calcium:Magnesium ratio	2.4:1	
Anion:Cation balance	35.5	

Breeding Stage (45-90 days)

Everyone knows the importance of efficient breeding. During this period the cow metabolism is concentrating on increasing milk production. Meanwhile we are concentrating on getting her bred back to keep her on cycle. To achieve both goals it will be required to significantly increase her energy inputs to support both milk production and reproduction. Most often during this period of lactation nearly all energy inputs are being utilized for milk production and little is left for body maintenance. During this period the ratio of calcium to phosphorus becomes critical to achieve conception. I recommend a 1.7:1, calcium-to-phosphorus ratio to enhance conception. The sample ration below is balanced at 1.8:1which should be adequate. A bad calcium-to-phosphorus balance is more often than not the cause for poor heats and conception. The following ration was balanced based on 60 days post calving producing 80 pounds of 3.8% fat, 3.0% protein milk and a Body Condition Score of 3.0.

Ration Name: Breeding Ration
Cow Profile Lactating Lactation # 2

Feed Name	As Fed	Dry Matter
Mixed Grass Pasture	90	24.3
Mixed hay	6	5.22
Organic Corn grain ground	17	14.96
Organic Molasses black strap	0.5	0.375
Limestone	0.32	0.317
Ca 23%:P 18%	0.125	0.121
Vit ADE Example	0.095	0.093
Salt-white	0.25	0.248
Mag oxide	0.063	0.062
Selenium 0.06%	0.015	0.015

Concentrations

Totals	114.368	45.711
Requirements	0	45.36
Difference	0	0.35

Forage NDF as % of NDF	92
Forage NDF as % of DM	34.95
NDF intake as % of BW	1.3
Total forage ADF (lbs)	9.12
2. Forage:Concentrate ratio	[65:35]
3. Calcium:Phosphorus ratio	1.8:1
Potassium:Calcium ratio	2.4:1
Calcium:Magnesium ratio	2.6:1
Anion:Cation balance	33.3

Peaking Stage (90-150 days)

During this stage the cow should be confirmed pregnant. If so, the calcium-to-phosphorus ratio will change to 2.0:1 to support milk production. Along with this, it is important to also maintain a higher energy input to support milk production. The forage-to-grain ratio will continue to be closer than in early and late lactation animals to achieve higher energy values. Energy added to the ration provides for body maintenance, milk production, embryo development and protein absorption. It is extremely important to maximize the cow's genetic potential to manufacture milk during this phase of lactation. The reason for this is simply that every pound of peak milk equals 200 pounds of milk over the course of the lactation. I am not justifying pushing or burning cows out, merely supporting their capabilities. I feel that grain levels should not be elevated more than 1-

This grain ration is balanced for a cow's stage of production.

2 pounds above the required amount to coincide with current production. The following ration was balanced based on 90 days post calving producing 100 pounds of 3.5% fat, 2.8% protein milk and a Body Condition Score of 2.5.

Ration Name: Peak Production Ration
Cow Profile Lactating Lactation # 2

Feed Name	As Fed	Dry Matter
Mixed Grass Pasture	110	29.7
Mixed hay	6	5.22
Organic Corn grain ground	20	17.6
Organic Molasses black strap	0.5	0.375
Limestone	0.43	0.426
Ca 23%:P 18%	0.1	0.097
Vit ADE Example	0.095	0.093
Salt-white	0.25	0.248
Mag oxide	0.095	0.093
Selenium 0.06%	0.02	0.02
Concentrations		
Totals	137.49	53.872
Requirements	0	52.395
Difference	0	1.48
Forage NDF as % of NDF	92.1	
Forage NDF as % of DM	35.17	
NDF intake as % of BW	1.5	
Total forage ADF (lbs)	10.74	
2. Forage:Concentrate ratio	[65:35]	
3. Calcium:Phosphorus ratio	1.9:1	
Potassium:Calcium ratio	2.4:1	
Calcium:Magnesium ratio	2.5:1	
Anion:Cation balance	34.1	

Persistent Production Stage (120-200 days)
Persistent production is the period when the cow is pregnant and will tend to plateau or level off in milk production. She will hit a comfort zone where she is capable of only so much milk and still build a calf without jeopardizing her own health. During this stage we should feed grain supplement to match her actual production based on pounds of milk. The calcium-to-phosphorus requirements will reduce to 1.8:1 ratio range; this is for maintenance and calf development. The energy-to-protein ratio will also remain stable during this period and the forage-to-grain ratio will remain fairly constant with a slight rise in forage. The following ration was balanced based on 150 days post calving producing 70 pounds of 3.6% fat, 3.0% protein milk and a Body Condition Score of 2.5.

Ration Name: Milk Persistence Ration
Cow Profile Lactating Lactation # 2

Feed Name	As Fed	Dry Matter
Mixed Grass Pasture	100	27
Mixed hay	6	5.22
Organic Corn grain ground	16	14.08
Organic Molasses black strap	0.5	0.375
Limestone	0.313	0.31
Ca 23%:P 18%	0.125	0.121
Vit ADE Example	0.095	0.093
Salt-white	0.313	0.31
Mag oxide	0.063	0.062
Selenium 0.06%	0.015	0.015
Concentrations		
Totals	123.424	47.586
Requirements	0	47.43
Difference	0	0.16
Forage NDF as % of NDF	93	
Forage NDF as % of DM	36.69	
NDF intake as % of BW	1.4	
Total forage ADF (lbs)	9.93	
2. Forage:Concentrate ratio	[68:32]	
3. Calcium:Phosphorus ratio	1.8:1	
Potassium:Calcium ratio	2.5:1	
Calcium:Magnesium ratio	2.6:1	
Anion:Cation balance	35.1	

Post Persistent Stage to Dry-off (200-305 days)
This is the "down the hill'" stage of lactation- don't fight it! When it is a cow's time to slow down there is very little a farmer or nutritionist can do to stop it. The cow's metabolism is dictating the milk production you will receive. The grain level should remain balanced with the actual milk production, not holding it up or leading it down. The cow is still building your future — the calf. There will be significant changes in the dietary ratios we have talked about. The calcium-to-phosphorus, forage-to-grain and protein-to-energy ratios will all remain constant in relation to each other, being reduced equally and gradually with the cow's lowering production. Keeping the cow profitable and building a healthy calf are the primary goals during this period of time. The following ration was balanced based on 250 days post calving producing 50 pounds of 3.7% fat, 3.0% protein milk and a Body Condition Score of 2.5.

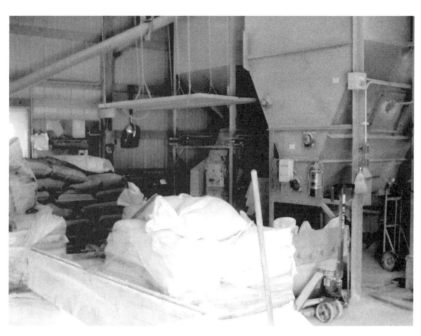

This feed mill only processes organic grains

Ration Name: Post Persistence Ration
Cow Profile Lactating Lactation # 2

Feed Name	As Fed	Dry Matter
Mixed Grass Pasture	80	21.6
Mixed hay	6	5.22
Organic Corn grain ground	15	13.2
Organic Molasses black strap	0.5	0.375
Limestone	0.2	0.198
Ca 23%:P 18%	0.095	0.092
Vit ADE Example	0.063	0.062
Salt-white	0.25	0.248
Mag oxide	0.045	0.044
Selenium 0.06%	0.015	0.015
Concentrations		
Totals	102.168	41.053
Requirements	0	41.79
Difference	0	-0.74
Forage NDF as % of NDF	92.2	
Forage NDF as % of DM	35.3	
NDF intake as % of BW	1.2	
Total forage ADF (lbs)	8.31	
2. Forage:Concentrate ratio	[65:35]	
3. Calcium:Phosphorus ratio	1.7:1	
Potassium:Calcium ratio	2.7:1	
Calcium:Magnesium ratio	2.5:1	
Anion:Cation balance	33.2	

Dry Cow Ration

This is the period between lactations when the cow is under the least amount of stress. Her missions in life are to finish developing the calf and build up her body reserves for another lactation. The calcium-to-phosphorus ratio drops to 1:4:1 and the forage-to-grain ratio rises to 70:30.

Ration Name: Dry Cow Ration
Cow Profile Dry Cow Lactation # 2

Feed Name	As Fed	Dry Matter
Mixed Grass Pasture	60	16.2
Mixed hay	4	3.48
Organic Corn grain ground	6	5.28
Organic Molasses black strap	0.5	0.375
Limestone	0	0
Ca 23%:P 18%	0.063	0.061
Vit ADE Example	0.063	0.062
Salt-white	0.063	0.062
Mag oxide	0.095	0.093
Selenium 0.06%	0.015	0.015
Concentrations		
Totals	70.799	25.628
Requirements	0	25.916
Difference	0	-0.29
Forage NDF as % of NDF	95.5	
Forage NDF as % of DM	41.56	
NDF intake as % of BW	0.8	
Total forage ADF (lbs)	6.08	
2. Forage:Concentrate ratio	[77:23]	
3. Calcium:Phosphorus ratio	1.4:1	
Potassium:Calcium ratio	3.6:1	
Calcium:Magnesium ratio	1.4:1	
Anion:Cation balance	39.3	

Chapter 3

Cow and Calf Management

The cycles of a milk cow bring her through a dry period and then the long lactation, when you are harvesting milk. It is understandable that your cow's system is going through very different physical changes in these different cycles. We will explain here things to consider as your cow goes through the dry period and moves into her lactation.

Calves are yet another issue. Successful calving and calf survival rates can make the difference between a successful and unsuccessful farm operation. Careful animal management is key to bringing these tender babies into the world and through their first year of development.

Dry and Lactating Cow Management
Dr. Paul Zettloff

Dry cow management starts in the last trimester of pregnancy when the cow is trailing way down in production. This is the period that the milk cow should be rebuilding her tissue reserves, the building blocks for all her systems. Elements for rebuilding include: trace and macro elements, amino acids, enzymes, peptides and others which are stored in the muscles and fat tissue reserves. The cow should be pretty much nutritionally restored at dry-off. The last thing you want is a thin, depleted, drained animal at dry-off. The six to eight week window when she is dry is not the time to rebuild her.

The last three months of lactation require skill and attention in nutrient management, as you do not want an animal to be hog fat either. Fat, over-conditioned cows at freshening will have calving difficulties and tend to have a rough time transitioning.

Following are four cardinal nutritional violations common in our dairy industry today. These have very negative effects on the dry and lactating cow.

Lack of free choice minerals

With the advent of the laptop computer and feed analysis, numbers can be crunched and a ration easily generated that prescribes all of the elements needed. The computer generated ration is definitely a wonderful tool, but it is an educated guess from a random sample. Mother Nature is bio-diverse and variable. Soils and feed can vary greatly, even in the same field. There should always be a safety net of free choice minerals available to balance out Mother Nature's (and your computer's!) variables.

Calcium and phosphorus are two big macro elements that should be offered free choice at all times. If you feed a lot of high calcium alfalfa, you will want more free choice phosphorus available. If you are on poor grass and corn silage, you will want both Ca and P. Salt is a trace mineral that should always be available. Look for a natural salt with no flowing agents, such as Redmond Natural Salt. This comes out of Utah and is available all over the U.S.

A mistake I often encounter is no available free choice mineral or salt for young stock. Growing musculo-skeletal systems need minerals. When I see a 500-pound calf chewing on the boards in a pen, I believe they need mineral and salt available free choice. When transitioning young stock onto salt and mineral, be careful initially as they may want to tank up on it and drink a lot of water, upsetting themselves. You may not be able to keep up with the resulting wet pen.

Minerals are an important component of dairy nutrition.

Overfeeding of grain and corn silage

Feeding a lot of corn silage to a dry cow is dynamite, as it will fatten up and round out your animal, stressing the liver with fatty infiltration of the liver cells. Corn silage is 50% grain and 50% poorly mineralized forage. Good feed for a dry cow is some long-stemmed, grassy hay, preferably low potassium hay.

High potassium in the dry cow ration is a cardinal no-no. The ideal is to have your potassium-to-calcium ratio as close to 1:1 as possible. A high potassium diet leads to three problems: alert downers at calving, udder edema and more left side displaced abomasums. When your potassium is 3 or over you are in dangerous territory. An exact 1:1 ratio is very hard to attain. If you can push your calcium up to 1.4 to 1.6 you have some decent feed, as trace minerals will follow calcium.

Everything starts with the soil. Soil balance directly affects the level and vitality of the cow's immune system. If you work with a soils program that balances the cations, calcium, magnesium, potassium and sometimes sodium, you will have better quality forage. The sulfur content also leads to better quality protein, as 10 percent of the amino acids of the protein building blocks contain sulfur. If you get a full-stemmed forage highly mineralized in the current proportion with sulfur levels in the 30 ppm on wet chemistry, you have a good dry cow feed and lactating cow ration.

Lack of knowledge of the immune system

There has been a good deal of information revealed about impact of the immune system on the cow's lactation. There are two natural dips in the immune system that occur like clockwork. Keep these two depressions in mind when you are assessing the strength of the immune system, and note them when you are setting up a vaccination program. These dips functionally correspond to two blips that are the inverse of the immune level we identify with cortisol. Cortisol comes from the endocrine system- mainly the adrenal gland.

Why do these two immune system depressions happen? Simply because the

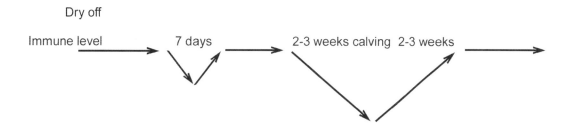

endocrine system, which runs all systems, is in a changing mode. At dry-off after about five days of a tight udder, the bovine changes from a lactating cow to a dry cow. Many hormones are changing to different levels. Before freshening, the system is starting to lactate. The bag is filling up and the pelvis is loosening up. These pre-calving changes are all controlled by hormones that operate in parts per billion. When drying off, always dry off and leave your cow for at least five, or even better, seven days before you touch the udder.

When have you have been told to vaccinate your dairy herd? I bet at dry-off and at calving. If there is ever a time in the immune cycle that is contraindicated, it is at these two dips — stay out of there with any vaccination. The dairy industry is over-vaccinated. We are doing more damage to the immune system by pounding it over and over with these new, highly antibody stimulating, slow releasing adjuvants. Vaccinate only for what is absolutely necessary and then look at using the old time nosodes. (A nosode (from nosos, the Greek word meaning disease) is a homeopathic preparation made from matter from a sick animal or person. Substances such as respiratory discharges or diseased tissues are used. The preparation, using alcohol, repeated dilution and succussion, essentially renders the substances harmless, while producing a powerful remedy. The use of nosodes for preventing disease has been employed in veterinary and human homeopathy for many years.)

Synthetic dry cow treatments in organic management systems are not allowed. When dry, only worry about cows or quarters with high cell counts or cows that have had a problem. Bring the problem cows in after they have dried off for 7 days, then strip them to see how they are doing. If they are full of mastitis and dead cells, you will need to get proactive. Here is where whey products shine. Let's look at one treatment option: A probiotic product (such as Crystal Whey or Impro) - 30ccs subcutaneous (SQ=under the skin) for three days, then skip a week, 30ccs SQ on the 10th day, skip a month and 30cc of the probiotic product SQ on day 40. During this time put her on an antioxidant and aloe pellets, 6 ounces a day, to elevate the immune system. The time to get proactive on mastitis is after dry-off. Work now to help the immune system. Kelp and other trace minerals may also be added during these windows of immune depression, and will be discussed later.

The best time to watch and evaluate the level of your herd's immune system is during the two to three weeks after calving. If you are getting retained afterbirth, mastitis, uterine infections, cows not eating enough or cows not coming into milk as expected, your cows may have low levels of immune function at calving.

Lack of pre-milking

Pre-milking is important in order to see what is in the udder. These big cows are different from cows of a hundred years ago. Modern cows produce vastly increased quantities of milk. When the tight, big, swollen udder presents itself at two or sometimes three weeks before calving, you wonder if there is a problem. Strip her out and start pre-milking her to see what is in there. If she has a problem, get proactive with garlic, aloe and other tools. If you must pre-milk her up to and including calving time, do it. Colostrum will be there because it is formed at calving. In fact, by pre-milking you have cleaned up the colostrum by getting rid of the dead cells and white blood cells that accumulate in the udder during the dry period. It has also been proven that pre-milking will lower the incidence of milk fever at calving because you have stimulated the parathyroid gland to get the calcium mobilized from the bone.

Somatic cell count

Herds that run high somatic cell counts have many areas to investigate. Look at the following areas: acidosis from too much grain, stray electrical currents, environmental and general cow sanitation, comfort, equipment and milking procedure. A general aid while one is trying to find the source of a high scc is to raise the overall immune level. There are three nutritional items that work well to do this:
1. Put the herd on two ounces of kelp per head per day. Kelp is loaded with trace minerals, the building blocks of all life. There are many companies that sell kelp. It is also sold in a salt-kelp mixture that can be available free choice. This is also a wonderful tool for young stock with pinkeye, ringworm or to boost overall health.

2. The second tool to take your herd system to a new level are the trace, botanical and neutraceutical items that can be added. There are many of these available. I am most familiar with Dairy Glow by Crystal Creek, Inc. These are excellent adjuncts to be used during times of stress or high performance. The best time to start these items is in the dry period. Remember to raise the level of everything before cortisol lowers the immune function during the two drops previously described. We are preparing for the two weeks before and three weeks after calving. These products have benefits for breeding as well. It is good to keep using them until the cow has come off her peak and is pregnant.
3. The third tool is aloe vera pellets in the feed. Aloe vera has the ability to circumvent cortisol's depressant function and engage the immune system to produce antibodies. This is a wonderful tool for high stress periods. The three-week window after calving when the cow is at her lowest for the year would be the time to feed three weeks of aloe pellets. If the herd develops a cough after a wet snow of a cold November wind, this is the time to uplift the system. Aloe is a useful aid to be used at the proper time to assist the cow to help herself.

The veterinary medicine of organic systems is not about bug killing, symptom smashing or injectible drugs. It is about health from the ground up. It is about a balanced natural life system using natural products to raise the animal's vitality for good production and longevity. We cannot bend Mother Nature. You have to learn how to crawl into her system by working in your own farm ecosystem where everything has a purpose.

A Self Fed Mineral Program
Dr. Richard Holliday

If you really want an education in mineral nutrition and want to give your animals a chance to balance their own mineral requirements, try the following program. At first, put out only small amounts and watch closely what they eat. Partition off your mineral feeder and provide the following in _separate_ compartments on a continuous, free choice basis.

- A mineral mix that is high in calcium with little or no phosphorus. You could use ground limestone (calcium carbonate) or oyster shell flour or combinations.

- A mineral mix that is high in phosphorus with little or no calcium. Providing calcium and phosphorus separately allows them to maintain the critical Ca/P ratio.

- Loose salt (not block salt), the more _un_refined the better.
- Kelp. This is a rich source of all trace minerals and iodine.
- Supplemental magnesium and potassium may not be necessary in all areas, but it does not hurt to make a feed-grade source available and see what happens. Magnesium oxide and magnesium sulfate are common sources. Both are relatively unpalatable. They can be mixed with salt to improve palatability so long as a separate source of plain salt is also available. An alternative is to provide dolomite limestone that contains Mg carbonate as well as Ca carbonate. In many areas, potassium is already adequate or excessive. Potassium chloride or potassium bicarbonate is commonly used in commercial mixes to supply this mineral.
- Sulfur is often deficient. Elemental sulfur can be provided free choice.
- Baking Soda or sodium bicarbonate free choice may be beneficial, especially if a lot of grain is being fed.
- If not already present in some of the other mixes, provide a source of vitamins A, D & E and some B vitamins.

Minerals are an important component of dairy nutrition.

More than likely, your animals will show a preference for one or two items, indicating a need. If your current ration is well balanced, they probably will not eat many additional minerals. Even so, leave it out for them and watch what happens to the consumption patterns over time when pasture conditions change or when feeding hay or grain from a new or different source. I have seen daily changes in mineral preferences for no discernible reason.

Avoid sudden changes to the ration. If your cows seem to grossly over-eat any one item, it may be prudent to partially limit that item for a week or so to let them catch-up gradually. If possible, avoid mineral mixes that are flavored to increase palatability. If you are already feeding a complete ration with minerals added, do not change the ration. Use this program as an add-on, free choice, monitoring system to let the animals tell you what they think of your ability as a nutritionist! This allows us to use our science and computers to at least get close to a balanced ration and still provide a way for the animals to fine-tune for their individual needs.

Managing Dairy Calves
Dan Leiterman

The following dairy calf management concepts are based on a combination of industry data, farmer comments I've received over the years and personal observations. All who raise calves have personal experiences that are of great value. My intent is to offer insights that may be useful when challenges arise, as can easily happen when nurturing such sensitive and important creatures as dairy calves. The focus for productive calf-raising should be on prevention rather than waiting for serious situations to develop where crisis management is needed and may not succeed. Sound and careful attention to the simple basics is good to remember when raising calves. The best calf raisers I have observed are people who care for the calves as though they were caring for their own children, where basic common sense is applied. It is important to do a very good job with the basics of calf raising, rather than being distracted or tempted to treat symptoms while ignoring root issues.

Remember, if a problem occurs, there is generally a way to prevent it — there is an answer. Good problem solving requires an open mind and a competent information network. Consistent, caring management, positive attitude and the realization that you have the ability to manage for prevention is key to maximizing calf health and reaching your production goals.

I will highlight here several tips that may be beneficial to your operation.

Birthing management

Organic calves thrive on pasture.

1. Clean dry maternity pen: Calves are like little sponges. There are several entry ports for serious infection to occur to this helpless creature. Use the 'dry knee' test with maternity pen bedding. If you can kneel on the bedding and come up with dry/clean knees it is probably good for birthing. Use plenty of dry, clean straw or comparable bedding.
2. Dip navel with 7% Iodine: One of the goals is to dry up the umbilical cord as fast as possible to prevent infection from entering the body. Do not use lower % iodine. Dip completely up to the and around the navel entry as soon after birth as possible.
3. Let the cow lick the calf: Research has shown that letting the cow lick the calf has many benefits to the calf, not the least of which is to improve nutrient absorption in the intestinal tract, i.e. absorption of antibodies from the colostrum.

4. Do not let the calf nurse the cow: Normally this is a very good way to deliver colostrum to the calf; however, there are several potential disease risks that need to be considered. For example, a major delivery mode for cryptosporidia infection (a common one cell parasite that can attack the digestive tract and can weaken calves) of calves is by letting the calf nurse a cow with a dirty udder. Cryptosporidia is common in the environment in fecal matter and a calf nursing the cow is a common infection route.

Colostrum feeding

One of the single most important management tools we have to ensure thrifty calves and reduce disease risk is to maximize the intake of highly functional antibodies from first milk colostrum. Proper feeding of good colostrum is critical to protecting the calf during the first 28 days of age while its active antibody production is not yet developed. Research has given the calf raising industry some new information to consider when managing for optimum colostrum delivery. First, it is important to understand that not all colostrum antibodies are functional and that not all mature cows have high levels of functional antibodies. Some antibodies may become non-functional due to improper handling (aging, heating, spoilage, storage, handling) as well as possibly due to excessive stress and nutritional deficiency to the cow during colostrum development. We are now learning that the difference between functional and nonfunctional antibodies is an important distinction. Years ago we used to think that heifer colostrum had lower levels of antibodies than mature cows. However today we realize that mature cows may have colostrum antibody levels as low or lower than heifers. The old rule of thumb of giving two quarts of colostrum may not provide adequate levels of functional antibodies to reach a target total blood protein test of 5g/dl or 2000mg/dl of true antibody.

Another consideration is that there are immune supporting small peptide fractions in colostrum other than antibodies that are beneficial to the calf and can stimulate immune response. These peptides are hard to measure even with blood tests. Because of these other immune supporting peptides, the target level of 5g/dl of total blood protein may not be the only or best indicator test for reaching optimum colostrum delivery. Also, using a colostra-meter (a bobber-like float put into colostrum to estimate anitbody level based on specific gravity, i.e the more antibodies the higher the bobber floats) is a crude tool to measure colostrum quality and lends no indication about antibody functionality. Variation of solid components in the milk may also affect colostrum's specific gravity. Until a better test is available, let's apply what we know and use the following standards or guidelines.

1. Feed One Gallon Of First Milk Colostrum: Use only first milkcolostrum for first feeding. Second milk colostrum has been shown to contain as much as 1/3 fewer antibodies.
2. Administer During The First 8 to 12 Hours Of Age: Preferably during the first 8 hours of age. The ability to absorb antibodies is greatly reduced after the first 8 hours of age. Most calf raisers would

Chapter 3 Cow and Calf Management

rather not orally drench (bag) colostrum to the calf. However, if it is necessary to achieve intake of one-gallon colostrum in the first 8 hours, it is worth it. Even if the calf does not want to eat at the next feeding, this practice can significantly reduce its risk of disease.

3) Use Johne's-Free Colostrum And Milk: A common mode of Johne's transfer to calves is through the colostrum and the cow's milk. Test to ensure the colostrum used is Johne's-free. When testing for Johne's I strongly recommend you use a fecal test rather than the Elisa blood test because the fecal test is much more accurate. Using Johne's-free colostrum and milk is important in reducing the risk of Johne's in the dairy herd so test well and protect your calves.

4) Consider Colostrum Supplements: If your herd has tested positive for Johne's it is important to use Johne's-free colostrum and milk. There may be a need to use a colostrum supplement to extend on farm colostrum supply. First generation commercial colostrum supplements were made from actual dried colostrum and are now becoming less popular in lieu of a higher performance second generation colostrum supplements based on serum antibodies. If a certified organic herd has tested positive for Johne's, organic certifying agents have the option to allow the use of colostrum supplements as well as non-medicated, all milk calf milk replacers to reduce the risk of Johne's.

> The NOP states: 205.603(c): If using a non-organic milk replacer: "Milk replacers without antibiotics, as emergency use only, no non-milk products or products from BST treated animals."

Young calf management

Calf comfort, reducing stress and controlling environmental challenges are very important basics in raising calves. Many calf raising disasters originate when these basic principles are violated. Ask yourself, would you put your children in the calf pens?

Comfortable calf housing is critical.

1. Clean, Fresh, Dry Air: Damp air breeds pathogens. Stale air stresses lungs. Be aware of ammonia build-up. Get on your hands and knees and take three deep breaths. If you are unwilling to get on your hands and knees, is it because it's dirty and unpleasant down there? Think about it. Once at the calf's level, breathe deeply and be observant to ammonia levels. If you find it unpleasant to breathe, there is significant stress to the calf's lungs and immune system. Ventilate the area with fresh air as soon as possible.
2. Avoid Drafts
3. Keep Calves Dry And Clean: Look at the knees of your calves. Are they clean, white and dry? They should be.

Calf nutrition

There is a very important point that I would like to emphasize here, and that is young calves (especially 1 to 21 day old calves) are still baby calves. Mother Nature has designed the young/baby calf to utilize milk as the best nutritional source as compared to other food groups such as grains or forages. The following discussion brings to light some basic facts about calf physiology, nutritional requirements and some common sense options for feeding a calf that are standard practice in other countries and are starting to gain recognition in this country. Calf survival and productive growth are directly tied to diet quality. It is critical to offer the very best nutrition in the best possible delivery manner in order to optimize a 'prevention' management style of raising calves.

Nutrient requirements

Let's look at basic nutrient requirements of a young calf as stated by the National Research Council (NRC), a non-profit organization that does basic research on livestock nutrition and sets nutritional standards for the livestock feed industry. (See figure 1.) It is important to remember this one basic concept in raising any young livestock, including calves. Calf weight gain directly correlates to calf survival and disease resistance. If the calf is not gaining weight, it is at immediate risk. How much weight should a calf be gaining? A moderate gain of 1 pound per day at 21 days of age is desirable, progressively giving the calf approximately 41 pounds of gain over the first 42 days of age.

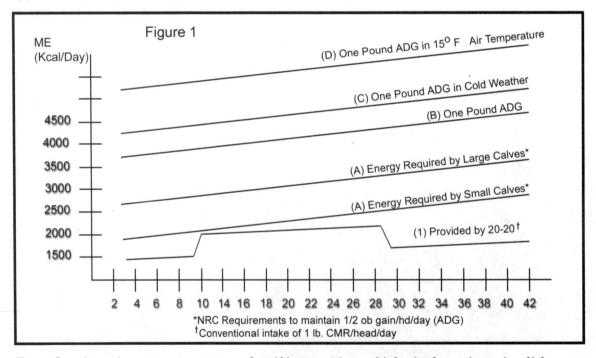

Figure One shows the energy requirements of a calf (approx. 85 pounds) for the first eight weeks of life.
Line 1: The calories provided by one pound of calf milk replacer powder fed per calf per day (equals two quarts of milk fed twice a day). Whole milk may provide a slightly higher energy level.
Line a: Calories required by the calf to gain ½ lb. per day.
Line b: Calories required by the calf to gain 1 lb. per day at 70° F. air temperature.
Line c: Calories required by the calf to gain 1 lb. per day at 45° F. air temperature.
Line d: Calories required by the calf to gain 1 lb. per day at 15° F. air temperature.

The two areas I would like you to focus on are the 7-10 day period and the 21 day period.

7 to 10 Day Old Calves

1. On about day seven, calves begin to starve if only receiving two quarts of milk twice a day with no calf feed being fed. Two quarts of milk twice a day only meets basic maintenance energy requirements for the seven-day-old calf to breathe, providing no weight gain, assuming no stress and at a 70°F air temperature. When fed this low level of milk nutrition, the calf is forced to eat other things in order to survive, i.e. dry calf starter feed, hay, bedding (which must be organic for the calf to be certified). On the surface this may seem to be a good thing when calves start to eat calf feed early; however, high calf feed intake during days 1 to 21 indicates a high risk situation. The calf's digestive tract is not developed well enough to properly digest some of the grains typically found in calf feed. Feeding grains to a very young calf can cause digestive stress, diarrhea/scours and other secondary challenges like pneumonia.

These calves are fed whole milk twice a day.

2. There are two grains in particular that prove difficult for calves to digest during days 1 to 21 of age: corn and soybean meal. Corn contains a starch that requires the calf to produce an enzyme called amylase in order to start digesting. Calves do not produce amylase very well (or in any significant level) during the first 3 weeks of age and find it hard to digestively deal with cornstarch. Other starch sources such as oat or barley starch are easier for the calf to digest. Soybean meal contains several compounds that are known to be anti-nutritional, which means these soy source compounds actually reduce nutrient digestion and nutrient absorption in the intestinal tract and can be irritating to the intestinal tissues. Other protein sources such as linseed meal have much lower levels of these anti-nutritional compounds. Also, heating the soybean products can neutralize many of the anti-nutritional compounds in soybeans, so roasted soybeans, extruded soybean meal or expeller meal are better options for the calf. Another potential concern in feeding soy products to young calves is the potential for estrogens in soybeans to negatively (and possibly permanently) affect reproductive performance of the calf during its life. Though inconclusive at this point, research is becoming more readily available on the potential hazards of feeding high estrogen feeds to young animals and should be watched as this topic develops. Remember too that all feed must be certified organic.

Meeting energy requirements for gain in 1 to 21 day old calves with grain based calf starter feeds may not be as beneficial to the calf when compared with meeting calf energy requirements by feeding higher levels of milk.

21 Day Old Calf
1. Note the level of energy two quarts of milk fed twice a day provides when compared with the calorie level required to grow one pound per day at 21 days of age. Then notice how energy levels need to increase as the air temperature gets colder. A 21-day-old calf could use two gallons of milk per day to meet energy requirements and a weight gain of one pound per day (a gain level that helps reduce the risk of disease). Since calves need significant levels of digestible calories at this young age to grow properly, and since their digestive tract is designed to digest milk, it seems only logical that we feed the proper levels of milk to meet their needs.
2. Proper weight gain has been associated with the development of other body systems in the calf, such as improved digestive development, including better rumen development. Rumen development relies on a high intake level of volatile fatty acids (Calories – which can come from milk.) If calves gain weight properly the tissues and organs of the body also grow. Young calf rumen development does not rely on early grain or forage intake, but rather depends on a high level of easily digestible calories like those found in milk. Digestive enzyme development also improves in young calves that gain weight properly. Feeding higher levels of milk during 1 to 21 days of age can help the calf utilize grain better when it is introduced after day 21 of age. Calf immune system development is also directly tied to weight gain — calves that gain weight properly are generally thriftier, more durable calves.

Tips for feeding higher levels of milk

If you are going to feed higher levels of milk to a calf, it is critically important to understand some basic rules to help reduce the risk of causing a digestive upset/scours and pneumonia. If the following considerations cannot be implemented, you may find it more comfortable to stay with the old-fashioned early grain approach.

1. <u>Optimum milk feeding temperature is approximately 100°F. to 105°F.</u> If milk temperature is too low at feeding time, the calf could misroute the milk into the undeveloped rumen where digestion is delayed and the calf is deprived of key nutrition when it needs it. The negative impact of this could be magnified in a cold weather environment because the calf may not be able to digest dietary fats properly and is less able to maintain body heat. In summer months, the misdirected curdled milk is less digestible and can propagate the growth of pathogens such as clostridium. Associated toxins may develop, threatening the calf's life. At the very least, the calf may scour and/or gain poorly.

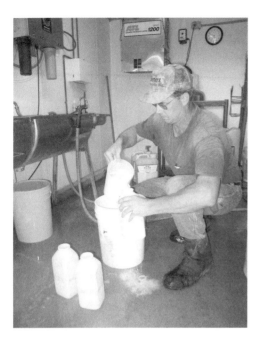

Most organic dairy farmers use whole milk to feed calves..

2. <u>Mixing temperature</u> for calf milk replacer powders should be with hot water (120°F to 140°F) to help disperse fats for better digestion. Also, milk replacers for young calves should be all milk (containing no soybean proteins and no products from rBST treated animals- NOP).
3. If significantly higher levels of milk are being fed it may be helpful to <u>feed three times a day, especially in winter.</u>
4. <u>Increase milk intake slowly</u>. Sudden increases in milkvolume may disrupt the digestive tract and cause scours. It may take a whole week to increase milk intake by ½ quart per <u>day</u>. It is possible to increase milk intake more quickly, however the rate of increase will depend on feeding temperature, calf health and milk quality.

Feeding strategy options

Calves fed gradually increasing amounts of milk tend to gain better, take to dry calf feed better when introduced, have less risk of digestive upset, have lower disease risk and can have higher calf feed intakes at eight weeks as compared to those fed a consistent milk ration. I know of Jersey producers that have their calves on <u>one gallon of Jersey milk twice a day</u> at 21 days of age and the calves are beautiful, thrifty and have very little health problems. <u>If higher levels of milk can be fed during days 1 to 21 then I recommend to start feeding dry calf feed at about 21 days of age.</u>

I am often asked when to start feeding baled hay because there is still the misperception that very young calves need scratch factor to develop the rumen. I recommend feeding baled hay after 6 weeks of age. It is harder for young calves to cope with fermented forages as compared to a good quality dry hay until they are older. I encourage you to build a solid foundation of preventative management by custom designing a feeding strategy for your operation based on these concepts.

Calf scours/diarrhea

Preventing calf scours is preferable to having to treat for them and the above tips can be very helpful in achieving a sound calf raising program based on prevention. With a good managerial attitude of taking responsibility for calf performance, we need to start with the assumption that most calf scours are preventable. Let's keep these guidelines in mind as we discuss the treating of calf scours.

Most Vulnerable Age For Calf Scours
1) Day 3 when calves change from colostrum milk to whole milk.
2) Day 7 to 10, when calves no longer can get enough energy from two quarts of milk per feeding and are forced to eat dry calf feed.
3) Post-weaned calves, especially due to possible coccidiosis infection.

Key Reasons Why Calves Scour
1) Abrupt change in diet.
2) Incorrect milk feeding temperature.
3) Overly aggressive increase of milk intake.
4) Starting on dry calf feed during day 1 to 21.
5) Environmental temperature stress, i.e. cold weather gut exhaustion due to shivering, hot weather dehydration.
6) Feeding poor quality milk, high somatic cell count milk, low grade calf milk replacers.
7) Feeding antibiotic laden milk.
8) Inconsistent feeding schedule.
9) Stress, i.e. moving, regrouping, crowding.
10) Feeding a diet containing anti-nutritional compounds, i.e. non heated soybean ingredients.
11) Infection of opportunistic pathogens in the digestive tract due to the above stresses.
12) Parasitic infestation, i.e. *Cryptosporidia, Coccidiosis, Giardia*.
13) Dirty pails and bottles.
14) Viral infections.
15) Feeding water too close to feeding the milk.
16) Incorrect dilution of the milk, i.e. too dilute or too concentrated can adversely impact intestinal osmolity.
17) Toxemia from moldy feed and/or forage.

Diagnosing pathogens related to scours

(see figure 2)

It is important to remember that the following information is only a general guide to interpreting calf scours. Having a lab analysis of cultured samples is the most accurate way to diagnosis pathogens associated with scours. It may be helpful to consider the calves' digestive tract in two general areas, the front half and the back half. It is not uncommon to have pathogens in the front half (mouth, throat, rumen, stomachs). However if the intestinal environment is disrupted due to any of the above listed issues, it may be possible for pathogens to invade the intestinal tract and cause or predispose a diarrhea condition. Once the intestinal environment and its supporting cast of beneficial bacteria are compromised, it is more likely that pathogens can become a problem. Be aware that it is possible to have an initial predisposing pathogen originate a problem (or be the first one affecting the intestinal tract), only to have other opportunistic pathogens follow as secondary infections to complicate the diagnosis and the decision as to which therapy is appropriate.

Figure 2

Dysentery Treatment Guide	
Physical Properties of Feces	**Primary Pathogen Indicated**
Yellow or white, runny, wet, messy rear end	E. coli
Grey, greenish, watery scours (calves less than 4 weeks old)	Cryptosporidia
Grey, greenish, brown, orangeish, watery scours (calves older than 4 weeks, up to older cows)	Coccidia
Grey, greenish, watery, foul smelling	Salmonella
Bloody scours (dark black or bright red)	Any of the above
Watery scours occurring at about day 7 of age, with a high death loss	Roto Virus
Dark, watery scours on older calves and cows	Winter Dysentery
Dark, watery scours on cows after freshening	Johne's
Watery, several or most cows in the herd	Toxemia from moldy/high spore count feed or pathogens

One example is when *Cryptosporidia* over-populates the intestinal tract, disrupts the environment and allows other pathogens like *E. coli* or *Salmonella* to invade. This is confusing when interpreting a lab culture which may show several pathogens. Which one is the real problem? The general guide listed in Figure 2 is also confusing. A calf may break out with a *Cryptosporidia* infection at midnight and have the typical grey/green/watery scours, but when the producer comes out to do chores in the morning the scours may now look yellow due to a secondary invasion of *E. coli*. Treating for *E. coli* in this case may only give temporary relief and not get to the root cause of *Cryptosporidia*. Consequently, it is important to combine several different observations to accurately diagnose the cause, apply proper therapy and subsequently to implement a plan of action for prevention.

There are a wide range of natural dietary supports (without using antibiotics and prohibited drugs) that may be beneficial to the calf during diarrhea challenges as well. Consult with your favorite natural vet or product distributor for recommendations.

Calves experiencing digestive stress may benefit from additional nutrition

Pneumonia and Respiratory Challenges
Pneumonia is generally a secondary problem that develops when the calf is already weakened by other challenges causing the calf's immune system to become overwhelmed, i.e. during diarrhea, stress, environmental or any of the previously listed challenges. Suffice it to say, calves at risk of respiratory challenges can be supported at many different levels with a variety of environmental and dietary adjustments. The organic producer has a wide range of supportive tools to address respiratory/pneumonia challenges with calves.

Weaning and Post Weaned Calves
It seems to me that weaning calves as early as 5 weeks of age seems a bit too young. It can be done; however, the question to ask is: Is it best for the calf? Conversely I tend to believe that weaning calves later than 9 or 10 weeks of age may be unnecessarily late. Weaning at 8 weeks of age seems about right, but other issues need to be taken into account other than just age of the calf or facility flow. I would recommend a combination of other criteria to affect the decision of when to wean.

1. Is the calf over 5 weeks of age?
2. Is the calf eating 2 pounds of dry calf starter feed?
3. Is the calf suffering from illness, i.e. scours, respiratory.
4. Is the body condition more than adequate to support an off-feed transition for the first post-wean week?
5. Will the post-wean environmental conditions be harsh, i.e. temperature extremes, over-crowding?
6. Is the calf eating hay properly?

Raising calves can be either a very rewarding experience or a very stressful situation depending on how well you have put the basics in place. The organic calf producer has a tremendous number of natural tools to address most all of the challenges you will encounter. Ask lots of questions and put your support team together so that when challenges do arise, you will be able to act in a timely and effective manner. Raising organic calves can be lots of fun.

Farmer Profile: Mark Eslinger - Boyd, WI
Jody Padgham

Mark Eslinger and his family have cut back their milking operation in the years since they went organic in 1993. "We just don't need the extra income anymore," Mark says. "With the farm paid off, and expenses low, 40 cows pay the bills and keep the work reasonable."

Mark rotationally grazes 90 of his 160 acres in north central Wisconsin. Now milking seasonally, he began transitioning to organic in 1990, one of the first farmers to do so in this very traditional dairy country. Over the years Mark has developed a good understanding of his land, cows and systems, which has spawned some very unique approaches to dairying.

Mark Eslinger assesses his pasture.

One system that has his neighbors scratching their heads is Mark's practice of leaving calves in with his dairy cows for up to three months. "These cows do a much better job of raising up their calves than I ever could," Mark says. "Labor is an issue with me — and since I don't need high production, and have low feed costs due to grazing and seasonal production, it works out in the long run for me." A neighbor at a recent pasture walk chimed in that he was starting to leave his calves on too. "The calves grow so fast, and look so good, that once you start you won't go back."

Three-month-old calves will drink up to 60 pounds of milk a day, which will put a good dent in anyone's milk check. But, Mark has never lost a calf to scours, has beautiful organic calves that get a good dollar at the sale barn and saves a lot on labor. He still milks twice a day, but production on those cows with calves drinking is significantly less than those later in their lactation. His overall herd average for his mixed herd is about 45 pounds per cow per day, with a somatic cell count average of about 150,000.

Mark will pull the calves out of the herd when the smallest is about 3 months old. He segregates the whole calf group and sets up a fencing system so that the mama cows can walk past the calves twice a day, but not get to them. Weaning takes about a week before everyone settles down. Calves get put on stored organic grain and hay and are raised up like beef calves till ready to be bred or sold.

Early on Mark did some culling to get cows freshening in a short window so he could dry the herd off for two months in the winter. Now about 75 percent of the Eslinger herd will freshen at the same time. "Put those cows on headed out grass and they will come right into estrus," he claims. Mark tries to get calves in April and May, as they are trouble free. Early or late calves will be the ones that have problems, he says. With a very low replacement level, Mark is still milking one or two 15-year-old cows. He is able to ship 10-20 very nice organic springing heifers each year, which helps to balance the lower milk check.

There are very few health problems at the Eslinger farm. Mark's theory is that the diversity of the pastures provides what the cows need to heal themselves. "They will find what they need out in the pasture," he claims. Any mastitis he sees is quickly cleared up by putting a calf on. Mark has found that the low stress of his grazing system means the cows are always clean and content, reducing health problems to almost nothing.

THE COW
The friendly cow all red and white,
I've with all my heart;
She gives me cream with all her might,
To eat with apple-tart,
She wanders lowing here and there,
And yet she cannot stray,
All in the pleasant open air,
The pleasant light if day,
And blows by all the winds that pass
And wet with all the showers,
She walks among the meadow grass
And eats the meadow flowers.
                by Robert Louis Stevenson

The Eslinger cows

Chapter 4

Organic Health Care

By now we should have you convinced that careful management of your dairy animals is essential to success on an organic farm. Without the "quick fixes" of antibiotics and other traditionally used drugs, building the immune system and preventative maintenance are keys to herd health. In the next several pages we will present you with options for health care in the organic herd that can be used in times of stress or when a health problem occurs.

Remember the answer to the question, "What can I use instead of (conventional product of choice) to treat my cows for (common disease or problem)?" In this chapter we will give you the answers for when Harriet Behar's solution ("Organic isn't about substituting approved products for chemical products. Organic is about managing your system for maximum soil, plant and animal health so that you don't need to use very many 'products' or off-farm inputs.") doesn't prevent or solve the problem.

Holistic Animal Husbandry
Dr. Richard Holliday

The success of the holistic approach requires a change in perspective and the development of a holistic outlook towards livestock management and disease control. A true holistic practitioner not only looks at the patient as an integrated unit but also views it in the context of the whole ecosystem in which it lives. In this regard, a sick animal is not only a patient to be treated but is also a symptom of a sick farm. Both patients need help. Any remedial action must include what is necessary for the immediate relief of the patient as well as a critical assessment of the long-term effects of the chosen therapy on the patient and the environment. Part of the treatment must also be the removal or reduction of predisposing factors. It is not as simple as merely substituting a "natural" alternate therapy for a "toxic" drug. The principles behind the success of holistic therapy go much deeper than the characteristics or source of the medication.

Conventional veterinary medicine is primarily concerned with the treatment of sick animals. Even if successful, the loss of life and production added to the cost of treatment makes this approach by far the more expensive.

Veterinarians also emphasize disease prevention. Herd health checks and vaccination programs fall into this category. As essential as these procedures are, the outlook is still towards preventing disease. Vaccinations may increase resistance against a specific organism but does little to elevate the animal's vitality to the health enhancement level. Typical of this category are herds or flocks where the animals are not really sick or showing symptoms but are not really well and productive either.

A third concept is that of health enhancement through holistic management. Everything possible is done to raise health and vitality to the highest level possible. All management practices are evaluated on the basis of their effects on the vitality of each animal in the herd. Strict attention is given to providing superlative nutrition. In so far as possible, all environmental stress factors are eliminated. Water is checked for nitrates or other toxins. Housing and ventilation are maintained at optimum levels. Any equipment with which the animals come in contact is properly maintained and adjusted. There are literally hundreds of other environmental factors that impact animal health and they all must be considered. When animals are maintained at a high level of vitality their resistance is much higher. Health enhancement is much more profitable than either treatment or prevention.

Adverse side effects from holistic medicines or procedures are uncommon. Those that do occur are mild and non-fatal, unless, of course, they are the result of gross negligence or ignorance.

Keep in mind that occasionally what appear to be adverse side effects may occur as part of the normal healing process. Many times the recovering patient will go through a 'healing crisis' before complete recovery takes place. During this 'crisis,' symptoms may intensify as the body begins to rid itself of toxins and healing progresses. An example of this is often seen when treating mastitis. As the udder begins to heal and the swelling recedes, the formerly dammed-off abnormal milk, pus and tissue debris is released. The sudden appearance of this "gargot" in previously normal looking milk causes the uninformed to think the mastitis is getting worse when in actuality it is only the body's way of cleansing itself.

Veterinary Tools for the Organic Herd
Dr. Paul Dettloff

The organic world has gone back to pre World War II archives and rediscovered tools available for animal health before the pharmaceutical world took over. In recent times we have discovered the working mechanisms of many of these effective old products, which are made direct from nature. Products that come from nature usually have very few side effects; due to their complexity we also rarely see resistance to natural antibacterial agents.

We can see the relative complexity between a manufactured product and a natural product by comparing the active antibacterial molecules found in a properly tinctured garlic and penicillin:

Known active constituents of penicillin are:
1) penicillin

Antibiotics are prohibited for organic livestock.

Known active constituents of garlic are:

1)	ajoene	15)	diallyl tetra sulfide
2)	allicain	16)	diallyl tri sulfide
3)	aliin	17)	dimethyl disulfide
4)	allixin	18)	dimethyl trisulfide
5)	allyn mercaptan	19)	dirpopyl disulfide
6)	allyl methyl thyosulfiante	20)	methyl ajeone
7)	allyl methyl trisulfide	21)	methyl allyl thiosulfinate
8)	allyl propyl disulfide	22)	propylene sulfide
9)	cysteine	23)	2-vinyl-4H-1
10)	diallyl disulfide	24)	3-tithiin
11)	diallyl hepta sulfide	25)	3-vinyl-4H-1
12)	diallyl hexa sulfide	26)	2-dithiin
13)	diallyl penta sulfide	27)	S-allyl cysteinew sulfoxide
14)	diallyl sulfide	28)	S-allyl mercapto

There are 9 veterinary tools that I regularly use in treating organic herds:
1) Tinctures
2) Homeopathy
3) Essential Oils
4) Aloe Products
5) Whey Products
6) Botanicals
7) Vitamins
8) Trace and Macro Elements
9) Probiotics

Tinctures

Tinctures are alcohol or glycerin based extracts of plants or minerals. The alcohol and glycerin absorb beneficial molecules that have medicinal properties in a concentrated form.

Garlic tincture is a mainstay antibacterial tool. Goldenseal is antibacterial as well. Eucalyptus is antibacterial and quite good against E.Coli bacteria. St. Johns Wort acts as an analgesic painkiller. Comfrey heals bones, osteoclasts speed up in the presence of comfrey. Arnica prevents bruising. Cauophyllum squeezes down the uterus. Echinacea tinctures kick the immune system and burdock root has very positive effect on liver health.

These are just a few of the tinctures that are currently available. We are seeing a rebirth in the use of these. There are some very good combinations of herbal tinctures when you get into the hormone replacement area. Tinctures are very valuable, safe and effective in the organic world. They are given in small amounts under the tongue or in the vulva/vagina if the animal is old enough to have had a heat.

Homeopathy

Homeopathy is the second tool that complements organic treatment. When properly applied, homeopathy works well by itself for a lot of deep-seated problems or it can be an adjunct to multiple therapies. There are some amazing results from homeopathy when applied by competent homeopaths.

Homeopathy was developed in Germany in the early 1800's by a gentleman by the name of Samuel Hahnemann. Homeopathy is based on the principle of "likes treating likes." For instance, if something gives you mild symptoms of malaria, it can be used as a treatment for malaria. There are hundreds of homeopathic treatments that have been described. They are taken orally in pill form. Some are very specific and work excellently on animals. A fine example is Apis mel, which helps prevent or treat swelling. Apis mel is derived from bee venom and bee antigen - likes treating likes. A second example is Rhus tox, which is derived from poison ivy. This is used to treat skin irritation and itching. A material medica is available that lists hundreds of these treatments. Although veterinary medicine has basically ignored this area in the academic world, there

Chapter 4 Organic Health Care

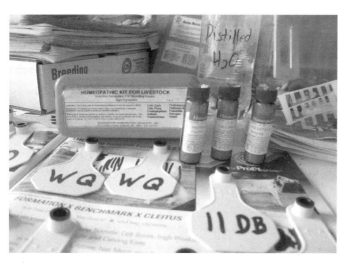

Homeopathy medicines are one tool for organic livestock health.

are some very good recent publications dealing with this topic.

Essential Oils

Essential oils are oils derived from plants and seeds. They are very potent and very concentrated, and can be extremely effective when used in the proper context. Essential oils are used in the human arena of aromatherapy with many ranging effects.

In the veterinary world a favorite use for essential oils is in liniments. For treating foot problems, melaleuca and eucalyptus work wonders on hairy warts. Respiratory problems can be helped with mullein leaf oils. Products with essential oils are not meant to be taken orally, they are usually applied topically.

Aloe Products

Aloe vera, the juice of a desert plant, is synonymous with "immune system stimulant". Cortisol, which is produced in the body during an acute disease or stress, lowers immune function markedly. Aloe vera has the ability to override the cortisol effect by stimulating increased immune function even when cortisol is present. This is paramount when treating acute illness. Aloe vera has great healing effects on skin and epidural tissue. There are 16 different positive effects that can be found from aloe vera — a mainstay and one of the first tools one should reach for in any veterinary situation.

Aloe vera comes in many forms: pellet for feeding, a liquid for drenching, a jelly, liniment and sprays for topical use. Safe and natural, this is one of the best bargains for organic dairy management. Be sure that the aloe vera used is certified organic and not diluted. An undiluted aloe liquid should contain 4000 to 5000 mucopoly saccharides per liter and should cost between 20-25 dollars per gallon. The seven-dollar gallon aloe is a diluted aloe mixture. You get what you pay for.

There is an excellent book on much of the recent research that has been done on aloe vera by Dr. Robert H. Davis titled *Aloe Vera: A Scientific Approach.*[5]

Whey Products

Whey products contain antigens and stimulate the immune system to produce antibodies. Some wheys are laden with antibodies that assist in healing many different systems in the body. Used properly, whey can have a dramatic effect on acute mastitis in cows and other animals that need an immune system boost. There are various kinds of whey products, some that are general and others that

---

[5] *Dr. Robert H. Davis "Aloe Vera: A Scientific Approach. Vantage Press, NY, 1997*

are formulated for specific uses. Whey has many good qualities. I am convinced that there are beneficial molecules yet to be discovered in whey.

Botanicals

Botanicals are one of the oldest known medicines. Botanicals are actual plant products that can be ingested orally, used in a poultice or brewed in a tea. A classic example is the common roadside plant mullein leaf. When dried it can be a food or a tea made for drenching. Mullein has tremendous expectorant qualities, which stimulate the motile cilia of the trachea to bring up fluid — the eyes will water, and copious nasal discharge will result when it is ingested. Mullein is also a tremendous aid in respiratory problems.

New uses of botanicals are being constantly developed for use in organic health management. Combinations have proven to be extremely effective—for instance, a good wormer is made of garlic, walnut hulls and leaves, elecampane root and wormwood.

Vitamins

Vitamins and antioxidants can be used to help pagocytes clean up cellular debris and foreign particles in a system. Vitamin C is a potent antioxidant that can be used in acute conditions. Tinctured rose hips and others can be used as antioxidants. Vitamin C, as an antioxidant, is underused in veterinary medicine.

Trace and Macro Elements

The reproductive system needs trace elements and enzymes to function. Calcium, phosphorus and all the other macro elements are vital to life. Organic farmers have learned to offer minerals free choice as a safety net when forages are available. These micro and macro elements must be recognized as a vital tool in healthy organic systems. Kelp is loaded with trace minerals and a wonderful tool for supplying some deficient items. Some of the first organic-focused soil and nutrition amendment companies introduced kelp to organic livestock systems. It quickly has become very popular.

Probiotics

Probiotics are probably one of the first tools available for organic systems, as they have been available in conventional dairy for some time. Probiotics are useful and necessary, but we find them less important as we concentrate on balanced soil, a balanced high forage diet and a proper running rumen. Probiotics can benefit a low pH, acidosis pushed, conventional cow more than a well functioning organic cow.

Take a moment to consider what the above listed tools are made of and where they come from. The answer for all of them is… Mother Nature. They are all molecules found in a natural state on this planet. When they enter an animal's system they are not a foreign, chemist-made molecule that is unrecognizable by the cell membrane of each metabolizing cell. The cell membrane controls all life. For this reason, side effects from natural remedies are very few.

Use of organic tools

Let us look at a logical approach in the organic treatment of some altered state of a dairy cow. Let us take for example an acute mastitis cow that has a toxic udder infected with a hemolytic *E. coli*, which is resistant to all the common antibiotics known. This is a common scenario. A cow milking 100 pounds a day, fresh 125 days, she has been working hard and her system is a little drained from this high production. In 12 hours she crashes and comes in with a hind quarter swollen double in size, no milk, some serum like product in the udder with only a few squirts, temperature is 106.4, heart rate over 100, rumen shut down, her blood stream has ruptured cellular debris, with foreign protein spilling into it from the udder and her lymphatics are plugged. She has just spiked her cortisol from her adrenal gland. She is a very sick cow and can die in 24 to 36 hours without help. What do you do?

You use as many tools on as many systems as you can, as quickly as you can to help her recover.

TREATMENT	TOOL	SYSTEM
IV Glucose 500cc	Instant energy	
Garlic, eucalyptus, goldenseal	Tincture	Circulatory
30 cc crystal whey SQ	Whey	Immune
300 cc aloe drench	Aloe vera	Immune
Phytolacta, bryonia or SSc	Homeopathy	Endocrine
Vitamin C	Vitamins	Lymphatic
Echinacea	Tincture	Immune
Liniment	Essential oils	Circulatory
Antioxidant blend	Tincture	Lymphatic
St. Johns wort	Tincture	Nervous (pain)
Strip out, strip, strip		Circulatory

Do you see what we are doing? We are not bug killing or symptom squashing her to make her look better tomorrow, we are using six different tools, some more than once, on five different systems of the body. We are trying to assist her in healing herself using her own devices.

You can expect her to appear to have a slower response visually than with conventional treatment, as we are not covering up any symptoms medically. But when systems respond, slow and sure, you will see fewer relapses. The higher the immune state you start with at the onset of the disease, the better the response. You can take this same approach with any acute problem. Each disease will have specific tools, but you will use the same approach, using the tools to affect the systems.

When we engage all these symptoms and speed up the immune system, building blocks for antibody production and enzyme production are needed (such as trace elements, amino acids, peptides and others.) Where are these? These are stored during good times in the muscle and fat of the body. The greater the tissue reserve, the more building blocks. These come from a balanced diet produced on good biologically active soil.

To be successful in organic medicine you have to see the entire system and help it stay balanced, nurturing it to optimum health. There is no magic bottle, only the use of a variety of organic tools on all the systems and a dose of good common sense.

Evaluation of Udder Health
Dr. Richard Holliday

One of the best ways to evaluate udder health is routine culturing (bacteriologic examination) of milk from any animal either showing mastitis or lower than normal milk production. Over time, these reports will allow you to arrive at a herd profile of the type infection present. Results interpreted on a herd basis rather than on an individual basis are of great value in managing the herd for maximum health.

Culture reports will not be meaningful if the samples are contaminated. If the germ that ends up in the tube comes from your hand or from a teat that was not properly cleaned, you could be misled into thinking it was the organism causing the problem. Contaminated samples are worse than no sample, so be sure and use extra clean procedures to collect samples. Results of culturing must always be correlated with symptoms. Remember, too, that if an animal has been treated with antibiotics in the previous 10-14 days any culture results will usually be negative.

Almost any bacteria can cause mastitis under certain circumstances, but most mastitis is caused by *Staphylococcus, Streptococcus, Escherichia coli* and *Enterobacter (Aerobacter) aerogenes*. It is not known why these bacteria become virulent at times, but stress is certainly a factor. If a high percentage of samples reveal the same pathogen, this is presumptive evidence of a cause and effect relationship between the pathogen and a specific environmental influence. These relationships are not absolute but they do provide clues about where to look first for answers. The following guidelines may help you match your problem to its cause.

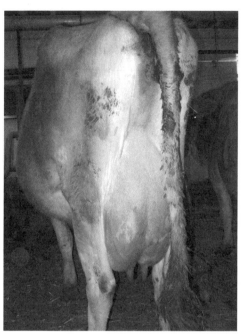

Monitoring udder health can prevent mastitis.

Staphylococcus

Staphylococcus bacteria have the ability to invade living tissue. Any physical damage, however slight, to body tissues opens the door for Staph infection. Of all the bacteria, Staph seems to have the greatest ability to quickly become resistant to antibiotics. Confirmation of this lies in the high incidence of post-surgical, antibiotic resistant, staph infections in humans. This condition is even known as "a hospital staph infection."

In dairy situations, two common causes of injured tissue that may lead to a Staph infection are improperly adjusted milking equipment and the use of irritating teat dips. Frostbite, stepped-on teats and other injuries may also be predisposing factors. Don't overlook the possibility of trauma just because you milk by hand. Hard stripping or milking entirely by stripping with wet hands can also damage the teat lining and open the door for Staph infection. If you have an ongoing problem with Staph infection, look for anything that causes injury to the teats or udder.

Streptococcus

Streptococci are not generally invasive but live on the surface of the udder tissue and in residual milk that is always present in varying amounts in the udder. Strep infection is generally seen when good milking techniques are lacking. It can also be associated with stray voltage or any other problem that interferes with milk let-down. If you have an ongoing problem with Strep infection, look first for anything that interferes with "let-down," "milk–out" or anything else that increases residual milk.

Escherichia

Escherichia coli — known as the manure bacteria — is found in all feces. Thus, mastitis caused by this bacteria is usually associated with unsanitary conditions. Some observations seem to indicate a higher incidence of *E. coli* when the ration contains excess protein, high levels of nitrates in feed or water or the addition of urea or other NPN's to the ration. If you have an ongoing problem with *E. coli* infection, look for anything that causes unsanitary conditions and check the water for nitrates and the feed for nitrates or excess protein.

Enterobacter

Enterobacter (formerly *Aerobacter*) *aerogenes* is often related to contaminated drinking water especially if animals have access to unsanitary water tanks, ponds, streams or puddles in the barnyard. If you have an ongoing problem with this infection, first check for the possibility of a contaminated water supply.

Some laboratories report *E. Coli, Enterobacter* and other Gram-negative simply as "coliforms." If a culture report lists any of these, I would strongly suggest culturing the water if you have not already done so. If the water is contaminated, remedial action should be taken at once.

Chapter 5

Stress and Handling

Stress is known to lower immune function and may be the primary factor that sets the stage for animal disease. There are many ways an animal can be exposed to stress in their life. Observation of your herd's typical behavior and times that behavior is interrupted is an excellent way to predict and discover stressful situations and circumstances. Learning proper animal handling techniques is essential, especially when you must transport your animals. Every time an animal is moved a stress reaction occurs, which can dramatically affect animal health and milk production. Proper euthanasia, when needed, is important in maintaining the integrity of your operation.

Some Thoughts on Stress
Dr. Richard Holliday

- Environmental or physical stress, such as faulty nutrition, bad water, lack of sanitation, poorly designed and maintained equipment, unsuitable habitat, etc. Good management has some influence on most of these but can not control all of them. For example, weather cannot be controlled but the effects can be mitigated with proper housing.

Fly tape lessens stress from flies.

Fans help reduce stress in a variety of ways.

- Physiological stress, usually associated with reproduction and lactation. We can minimize some of the effects of this type of stress, but we can not totally eliminate it.
- Psychological stress may occur when weaning or changing groups, establishing a new "pecking order," etc. This type of stress can be held to an acceptable level with good management.

All animals vary in their ability to accommodate stress. Some differences are due to inheritance, species, breed and sex. Others are associated with the individual's life history of health and disease. Older animals do not accommodate stress as well as younger ones do. A young animal that suffers an episode of severe scours/pneumonia may survive, grow and appear thrifty even though some irreversible damage to heart, lungs and intestinal lining may forever impair its ability to pump blood and absorb oxygen and nutrients. Under stress this animal will probably show earlier and more severe symptoms than others in the same group that did not go through the sickness

Stresses are cumulative. A small stress has a greater effect in an animal already carrying a big stress load than it has in another relatively stress free animal.

Zinc
Dr. Richard Holliday

Stress (including parturition) appears to increase the zinc requirement of animals.

Zinc is required for the incorporation of cystine into keratin and thus plays an important role in maintaining hoof, horn and skin integrity.

Zinc plays an important role in wound healing, immune function and disease resistance. Some studies indicate that the first symptoms of a zinc deficiency are a decrease in immune function and a decrease in feed conversion.

Zinc plays a role in vitamin A transport and utilization and appears to play a role in vitamin E absorption. Reproductive performance after parturition improves with both zinc and vitamin E supplementation in late pregnancy.

High calcium and iron intake (including Ca and Fe in water) will increase the zinc requirement.

Deficiency symptoms may include general listlessness, poor growth, stiff joints and unthrifty appearance, hair loss, general dermatitis of head and neck and failure of wounds to heal properly.

Animal Handling
Tamiko Thomas

Stress in a dairy cow can negatively affect health and production and so understanding low stress handling is very important. Handling interactions ultimately determine the relationship that develops between producers and their animals. To develop a positive relationship the animals must be handled in a humane manner, which requires a basic understanding of cow behavior and how it is affected by previous experiences, the environment, and essentially how they view the world. Optimizing handling is particularly important for lactating dairy cows because they are intensively handled throughout their lifetime.

Cattle Behavior

An animal's reaction to handling (and other external factors) will be affected by how it is perceived through the animal's sense of sight, taste, touch, hearing and smell. Vision in particular has been shown to be the dominant sense in many situations and to be responsible for approximately 50 percent of a cow's total sensory information. Having eyes positioned on the side of the head affords cattle wide angle vision, around 330°. They do, though, have a blind spot directly behind them, so when approaching from the back it is a good idea to call out gently, to keep them from starting. Cattle can see in color, especially at long wavelengths (yellow, orange, and red).[6]

Hearing is also of particular concern during handling. Cattle have a greater audible range of frequencies than humans and therefore may be disturbed by high-pitched sounds that are not audible to humans.

Dairy cattle are social animals that function within a herd structure. They will follow a leader to and from the pasture or milking parlor. This social nature and herd instinct means that isolation from herd mates is

Cows naturally follow a leader. A wide lane allows freedom of movement.

---

[6]Phillips. C.J.C. 1993. Cattle Behaviour. Farming Press Books. Ipswich, UK

stressful. In response to isolation combined with a new situation (which is frequently stressful) one study found that cows showed signs of acute stress. Their heart rates were higher and they gave less milk than controls due to reduced secretion of oxytocin, the hormone responsible for letdown, and increased milk retention.[7] Agitation due to separation from the herd can also make handling the animal more dangerous. Behavior during handling can also be affected by previous experiences. Animals remember frightening or painful experiences for months.

Human behavior

There is ample evidence that fear of humans due to inappropriate handling, such as using prods, yelling, hitting and making sudden jerky movements towards the animal, lowers productivity in dairy cows. In one study, fear of humans accounted for 20 percent of the variation in milk yield between farms.[8] Another study found that an aversive handler, who over a 3-day period hit cows or intermittently used a battery-operated prodder, reduced short-term milk yield by 10 percent and increased residual milk by 71 percent.[9]

Young animals also respond poorly to aversive treatment. A study of calves showed that they learned to discriminate between people based on their previous experience, approaching those who handled them positively and avoiding those who handled them aversively.[10]

Negative handling can further affect your bottom line by resulting in unfavorable behavior patterns developed over time due to stressful handling. More time and effort may be required to work with such cattle and this will undoubtedly add to labor costs.

The aversive handling described above is entirely preventable. Human frustration which can lead to abusive handling can be reduced by using an animal's natural behavior patterns to make handling and restraint more humane and efficient. Calm handling means calm animals which are easier to handle and move than excited animals. Proper facility design and clear paths

---

[7] Rushen, J., Munksgaard, P.G. Marnet, and A.M. de Passillé. 2001. *Human contact and the effects of acute stress on cows at milking.* Appl. Anim. Behav. Sci. 73:1-14
[8] Breuer, K., P.H. Hemsworth, and G.J. Coleman. 2003. *The effect of positive or negative handling on the behavioural and physiological responses of nonlactating heifers.* Appl. Anim. Behav. Sci. 84:3-22
[9] Hemsworth, P.H. and G.J. Coleman. 1998. *Humane-Livestock Interactions: The Stockperson and the Productivity and Welfare of Intensively Farmed Animals.* Cab International Oxon, UK
[10] A.M. de Passillé, J. Rushen, J. Ladewig, C. Petherick. 1996. *Dairy calves' discrimination of people based on previous handling.* J. Anim. Sci. 74:969-974

also facilitate movement. Care should be taken to avoid shadows, puddles of water, obstacles such as drains or metal pipes and flapping objects that make animals balk.

However, not all negative experiences are avoidable and some routine husbandry procedures can be aversive. The effect these procedures have on the animals depends on how aversive the procedure is and whether people are associated with that aversion. Rewarding experiences such as offering preferred feed or even positive handling around the time of the procedure may lessen the aversion to the procedure. This reduces the chances that animals associate the punishment of the procedure with humans. Furthermore, training animals to enter restraint, the milk parlor, etc. without doing anything aversive but instead providing food rewards can facilitate future procedures. Routine contact with humans from birth on, including regular gentle treatment, will reduce fear and flight distance and make observation and treatment easier. It will also enhance animal well-being and productivity- results that are important in themselves.

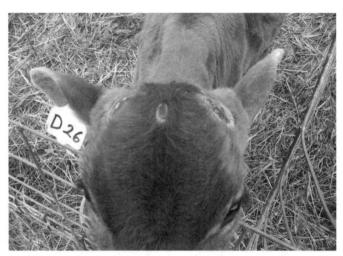

Physical alterations, such as dehorning, must be done at a young age.

Dairy animal transportation

While often necessary, transportation is inherently stressful to animals and has been shown to cause negative changes in physiology and behavior. Minimization of this stress is not only better for the animal, but also translates into significant benefits and monetary advantages for the farmer, whatever the reason for transport. It is important to consider carefully the various factors involved in the transport of dairy animals in order to minimize stress and prevent injury or death. Elements that must be considered and addressed include:

1. fitness of the animal to be transported;
2. facilities and handling techniques;
3. transport conditions and suitability of the vehicle.

Condition of the Animal to be Transported

Cattle must be fit before being transported--it is impossible to assure good animal welfare during transport otherwise. Examples of animals that should not be transported are cattle that are injured, severely lame, weak or emaciated. Animals suffering from these conditions are more likely to fall down in a truck and become non-ambulatory. Non-ambulatory cattle "downers" should be euthanized on farm. One estimate shows that dairy cows make up 75-80 percent of the non-ambulatory animals arriving at stockyards.[11] As of 2003 non-ambulatory cows are no longer allowed to be shipped. See box below.

---

[11] Grandin, T. 2000. *Perspectives on transportation issues: The importance of having physically fit cattle and pigs.* Presented at the Transportation Symposium at the American Association of Animal Science, July 2000. J. of Anim. Sci. 79 (E Suppl.)

> **BSE**
>
> On December 30th, 2003 non-ambulatory cattle were banned from the human food supply. This prudent move was made by the U.S. Department of Agriculture in response to the discovery of a Washington State cow suffering from Bovine Spongiform Encephalopathy (BSE) or mad cow disease. This regulation will help protect food safety and by removing the incentive for shipping live downed animals reduce animal suffering.
>
> BSE is thought to be caused by feeding material from ruminants infected with the disease to other ruminants. Organic dairy farmers are taking important steps to minimize risks of BSE and other diseases by not feeding their livestock feed containing urea, manure, or mammalian or poultry slaughter by-products, in conformity with the National Organic Program.

Potential food safety problems result from the way in which cattle are handled prior to slaughter and processed after slaughter. During processing meat from culled dairy cows is often mixed together and made into products such as hamburger, thereby potentially amplifying any contamination. Dairy animals should be transported before they become infirm.

Other transportation considerations are whether the animal is pregnant or has been recently treated with antibiotics. If a cow is pregnant it is recommended that she not be transported within at least 2 weeks of freshening. Antibiotic withdrawal times should be checked before animals are culled or sold.

The transport of dairy calves is also of great concern as they are particularly vulnerable to transport stresses. Calves should not be ill or have diarrhea, and they must have received adequate amounts of colostrum as discussed in the calf care chapter. Furthermore, their coat and navel cord should be dry and they ought to be able to walk easily without assistance before being transported. While it is common in the United States for bull calves to be transported at 1-2 days of age, calves are not fit for transport at this age. Calves are more likely to become non-ambulatory during marketing and transportation if they are less than two days old. Non-ambulatory calves are difficult to handle properly. Inappropriate handling such as throwing, dropping and excessive prodding has been reported. Long distance transport of young calves should be avoided. It is recommended that calves be a minimum of five days old before transport. In Europe and Canada they must be at least a week old before they are shipped.

Facilities and Handling Techniques for Loading and Unloading
Dairy cattle should be loaded and unloaded for transit in a way that minimizes stress and anxiety. Adequate numbers of properly trained caretakers are needed to accomplish the task humanely and safely. Loading areas should be situated near hospital pens and roads and be reachable in all kinds of weather. The pathway onto the vehicle should be clear and the number of directional changes an animal can take must be minimized. Ramps should not exceed a 25 degree angle. They should have non-slip flooring, solid sides, and there should be no holes between the ramp and truck.

Transport Conditions
The sheer number of animals that are transported each year makes it clear why transport accidents are one of the most common disasters that livestock owners will encounter. In order to reduce this likelihood the vehicles used to haul livestock must be properly maintained and designed. They should have sturdy sides and partitions when appropriate that are high enough to prevent livestock from jumping, falling or being pushed out. If transporters are hired they must have knowledge about animal well-being concerns and have species appropriate equipment. They should also have contingency plans and information on who to contact in case of an emergency. The cattle owner should provide instructions as to the care expected during transit.

Studies indicate that of all the various elements associated with transport (including handling, mixing of animals etc.), extended transport time appears to be the most stressful. Therefore transit time should be reduced as much as possible. On long journeys, cattle exhibit not only behavioral and physiological evidence of reduced welfare, but also greater risk of injury or death. Routes and time of travel should be planned so as to reduce delays due to traffic, road repairs, ferries, etc.

Social regrouping should also be minimized as it is a potential stressor. Animals that are likely to fight with each other should be segregated while in transit in the same vehicle. Animals that are of substantially differing weights or ages, except for a cow and nursing calf, must be segregated.

It is important to maintain appropriate stocking densities on transport vehicles. Animals must have enough floor space and adequate head room to allow for sufficient ventilation and for normal movement and resting positions. The most common hazard is overloading the vehicle, resulting in the last few cattle having to be driven forcefully aboard. Individual animals transported in overcrowded conditions may be unable to get up should they lose their footing. They may be trampled or injured.

Canada's Recommended Code of Practice for the Care and Handling of Farm Animals[12] during transportation identifies the following animal behaviors and conditions as warning signs of problems:

> *Overcrowding* Results in loads that will not "settle" but instead continue to be noisy for a prolonged period of time. Animals involuntarily lie down and are then unable to get up.
>
> *Overheating* All species will pant when overheated. Animals standing with neck extended with open mouthed breathing indicate a dangerous situation. Cattle should remain dry during transport and are probably overheating if wet.
>
> *Cold exposure* All species will eat available bedding when cold stressed. Fluid will freeze to the face or nostrils of cattle and under extreme cold stress cattle will shiver.

Animals should be provided with bedding for comfort and protected from temperature extremes during transport. In hot humid weather cattle require more ventilation during transport to prevent dangerous levels of heat buildup. This situation is most easily addressed by decreasing stocking density. Transporting at the times of day when the heat is not as bad is also appropriate. During cold weather an increase in the amount of bedding is necessary. It also helps to cover some ventilation openings to protect the animal from wind chill while still providing adequate ventilation.

Driving style and the quality of the roads are major factors determining the well-being of cattle during road transport. The importance of repeated examination of the livestock and of careful driving cannot be overemphasized. Careful driving means that drivers should avoid sudden starts and stops and negotiate turns in as smooth a manner as possible so that the cattle don't lose their balance. It is extremely hazardous for large animals to go down during transport.

Choosing transport routes with smooth roads is also important. One study on the issue found that the heart rate of beef heifers was lower, indicating reduced stress, when the vehicle was traveling smoothly on highways, compared with rougher country roads with numerous intersections.[13]

Euthanasia

Euthanasia is the act of inducing a painless death. The word itself is derived from the ancient Greek words eu and thanatos, which respectively mean good and death. It is important that irrevocably injured or sick cattle that are suffering receive a "good death." Such a death must be painless and requires a procedure that induces rapid unconsciousness, followed by respiratory or cardiac arrest and ultimately loss of brain function.

---

[12] *Canadian Agri-Food Research Council. 2001. Code of Practice for the Care and Handling of Farm Animals- Transportation. Website: www.crac-crac.ca/english/codes_of_practice*

[13] *Grandin, T. Cattle Transport. In: Grandin, T (ed) Livestock Handling and Transport 2nd edition. CAB International Oxon, Uk*

Persons who routinely work with livestock need to be properly trained to carry out emergency euthanasia. They should be skilled and use only techniques approved by the American Veterinary Medical Association. Acceptable methods are: barbiturate overdose, gunshot, or captive bolt. The first method involves a restricted drug to which only licensed veterinarians have access. The second method requires extreme caution because it uses free bullets which can ricochet injuring the operator or others. The firearm should be held perpendicular to the front of the skull to prevent ricochet and 2 to 10 inches away from the animal's forehead. Accurate placement of the bullet entry point is crucial. For adult cows the point of entry should be at the intersection of two imaginary lines drawn from the middle of each eye to the center of the base of the opposite horn. Mature bulls require special consideration because they may have a hard, thick frontal bone, often covered in dense, matted hair. A firearm using a 9mm shot is recommended.

Captive bolts can be either penetrating or non-penetrating. The penetrating version propels a steel bolt into the animal's brain. It kills by concussion and physical destruction of the brain similar to a free bullet. The non-penetrating version only stuns the animal and should not be used for euthanasia of livestock in field situations. After use of either the penetrating or non-penetrating captive bolt gun another procedure, such as exsanguination, is required. Exsanguination is the draining of an animal's blood, achieved by severing large blood vessels such as the carotid arteries, and it should be done within 30 seconds of stunning. In either case, the captive bolt must be pressed firmly against the animal's forehead and therefore restraint or prior sedation may be necessary to ensure proper placement. The point of contact for the captive bolt is the same as for the firearm. Calves can be euthanized without problems with a penetrating captive bolt gun. Firearms and captive bolts must be cleaned and maintained in order to ensure that they are effective. Any instructions for them should be followed closely.

After euthanasia dairy cattle should be checked for vital signs. The absence of rhythmic breathing, heartbeat, and corneal reflex for more than 5 minutes should be used to confirm death. Then the body should be disposed of properly and promptly in order to address public health and disease concerns. The carcass can be sent for rendering (unless a chemical was used for euthanasia) or disposed of by other appropriate means such as burial or composting. These methods must be done in accordance with applicable ordinances.

Chapter 6

Farm Milk Quality

Louise Hemstead

Marketing organic milk will generally bring you a premium in the marketplace, but this will not be the case unless your milk quality is very high. Many producers have scratched their heads over high somatic cell counts, wondering what pipe may be dirty or what cow has a touch of mastitis. Being careful at all levels of production, from cow health and cleanliness to parlor cleanout procedures is essential to maintaining a consistent and high level of milk quality. In this chapter we will walk you through the issues involved in maintaining the highest quality milk.

Quality has many definitions. According to the Webster Dictionary quality is: "An inherent or distinguishing characteristic; a property." To the consumer quality is about receiving what they expect from a purchased product. To a food connoisseur, quality is the piquant and delicate aroma of a finely aged cheese. In organic production quality begins at the farm. From feed to cow to milk there are many steps which impact quality.

Organic dairy manufacturers are very concerned about the impact of raw milk quality on finished product quality. The organic industry itself has a lower limit for somatic cells then conventional dairy, and many of the marketers for organic milk have established premiums and deduct for bacteria levels in the milk. Though consistent with the conventional milk industry, it is perhaps more critical for organic milk to be high quality in order to maintain the price point of organic products in the marketplace.

Today organic production is prevalent in regions of the country hospitable to dairy production, while organic dairy consumption is nationwide. As a result, we are shipping products farther and they are staying in the chain of distribution much longer then they did when organics were in their infancy. Despite the use of state of the art processing, it is apparent that excellent raw milk quality is the vital and extremely important component for production of top quality organic dairy products.

At the same time, today's consumers are prone to abuse their groceries in the process of transport from store to home. A somewhat cavalier attitude is taken regarding the temperatures of home refrigerators, and it is common to find consumers who leave milk resting at room temperature for an hour or two through out the supper-making process. This same consumer is paying top dollar for these products and may not return to purchase again if there is any perception of loss of quality. These conditions put pressure on the entire milk quality process.

There are several components that may lead to the perception of milk quality. When the ultimate consumer drinks a glass of milk, the flavor should be clean, slightly sweet and refreshing. This is not accidental or the result of idle chance. The flavor at the table has its roots in the feed, the cow and the barn/milking parlor that generate the milk.

To begin, let's revisit the components of milk. An ingredient statement on a product "Milk" would read as follows: Water, Sugar, Butterfat, Protein, Minerals.

Water is the main component of milk — obviously it is a fluid liquid product. Lactose is milk sugar. It is comprised of two sugars, glucose and galactose. Lactose in its natural state is not nearly as sweet as either of its components separately. In the cheesemaking process lactose is converted to lactic acid.

Butterfat molecules in their natural state are irregular and lighter then the other components in milk. Fat can be attacked by specific bacteria causing "lipolysis." This results in a rancid flavor in the milk.

The protein in milk is primarily comprised of casein. Casein is critical to the production of cheese. It is the casein and fat reactions to cultures that bind to create cheese. Another part of milk protein is commonly known as whey proteins. Whey proteins are less than 20 percent of the milk protein. An increase in somatic cells causes a related increase in whey protein and decrease in casein in the milk.

Milk is comprised of a small percentage of minerals, including calcium.

To a great extent, the breed of cow drives the composition of the milk. The following chart shows the variations of three common breeds.

	Water	Lactose	Fat	Protein	Minerals/Ash/Salts
Holstein	87.74	4.89	3.40	3.30	0.67
Guernsey	85.37	4.97	5.03	3.89	0.74
Jersey	85.46	5.00	5.05	3.89	0.70

Dairy Manufacturing Programs Dept of Food Science Cooperative Extension Service, University of WI - Madison Joppa B0131

Over the years careful genetic selection has allowed the producer to select genetics for the purpose of increasing butterfat and protein, the two most highly praised components of milk. In addition to genetics the ration fed to the cow has a direct impact on butterfat and milk composition. Dramatic changes in feed not only impact the cows' appetite, but also her yield of protein and butterfat. In some instances undesirable flavors can even be imparted directly to the milk.

Feed

The ration fed to cattle has a direct correlation to the flavor of the milk. Common odors or flavors that are transferred into dairy products include: malt, garlic, fresh grass, sulfur, alfalfa and weeds. These odors can permeate the herd, milk and ultimately, the tanker of milk. Raw receiving standards call for the rejection of tankers with these off flavors. Spring and fall are the seasons of the most common complaints by consumers relating to the flavor of milk that is traced to the farm. Both silage and fresh grass can impact dramatic flavor changes in the herd. Avoidance of dramatic ration changes minimizes the impact of flavors in the milk. Some consumers prefer the taste of fresh grass milk, and some farmers have found a niche in marketing "grass fed" dairy products.

A more insidious concern with feed is the presence of mold. Mold not only affects the taste and quality of the milk, but also the health of the cow. Penicillium is a mold. It is rare, but it is possible for moldy feed to contaminate a herd of cows and cause a positive antibiotic test result, and thus a challenge to organic status. While this is VERY rare, it does happen — should producers suspect they have a mold problem, they should not feed this contaminated feedstuff to their milking or meat animals.

Another challenge with feed is Aflatoxin M1. Aflatoxin M1 is a Mycotoxin which is carcinogenic. Mycotoxins typically appear in drought years with late rain. Corn that is stunted from maturity does not dry evenly and mycotoxins are often present in the silage. The cow is not an effective filter and the mycotoxins pass through her system into the milk. Several states now screen milk for mycotoxins. If a herd is identified with a positive Aflatoxin M1 screening test, the milk is withdrawn from the market until such time that the test is negative.

The cow

Milk is free of bacteria at the moment it comes from the secretion cells in the mammary glands. However, tissues inside the udder commonly harbor some bacteria. As milk travels through these tissues to the teat end it may become contaminated with bacteria. As a result, milk taken from the udder in a sanitary way to prevent external contamination can and commonly does contain from several hundred to several thousand bacteria per milliliter. Generally these bacteria are likely to be harmless to humans, however given the right conditions to grow, they can spoil milk through the production of acids and flavor compounds. We will later discuss the potential post cow contamination of milk.

Historically Tuberculosis, *Brucella abortus* (Undulant Fever/Bangs), *Salmonella typhi* (Typhoid Fever); *Cornybacterium diphtheria* (Diphtheria) and *Streptococcus pyogense* (Scarlet Fever/Septic Sore Throat) were all tied to the consumption of non-pasteurized milk and milk products. These diseases thankfully have been largely erradicated due to the widespread use of pasteurization.

Today other bacteria challenge the dairy industry: *Streptococcus agalictia, Staphylococcus aureus, E. coli, Listeria monocytogenes*, and *Mycobacterium paratuberculosis* all cause diseases which impact the dairy industry and have a relationship to human illness.

Streptococcus agalactiae, commonly known as Strep-ag, and *Staphylococcus aureus* are the most commonly discussed causes of mastitis. Not only do these bacteria cause mastitis outbreaks in dairy herds, but when consumed cause illness in people.

Clean bedding helps prevent bacterial infection.

For years it was thought that Strep-ag did not cause illness in people, but this assumption has been proven incorrect. People who work with cattle daily often become carriers for the bacterium. In the right circumstances the bacterium can cause pyelonephritis (deterioration of kidneys), pneumonia, meningitis and septic arthritis. Staph is more commonly known as the bacterium that causes food poisoning. Outbreaks in people are characterized by diarrhea and vomiting.

Somatic Cells
The NOP does not state a maximum allowable somatic cell count. However, a high scc is always an issue of concern, and individual marketing agencies will often have a maximum allowable scc level. For example, CROPP/Organic Valley requires that a herd maintain somatic cell counts at a level of less then 400,000 per milliliter.

The presence of somatic cells indicates a bacterial infection in the udder or the cow herself. Somatic cells are white blood cells, which rush to the body's aid to repair points of injuries. It is a cow's way of healing herself. However, when milk is contaminated with a significant level of somatic cells there is a direct impact to the relationship of the proteins in the milk. While the legal requirement for somatic cells is <750,000, the dairy industry has known that the presence of somatic cells decrease yields and causes organoleptic (flavor) challenges in product. It is standard for producers to be offered premiums to reduce the somatic level in milk. A gallon of milk on the grocery shelf today will generally show average somatic cell levels well below 200,000 per milliliter.

The primary changes in milk quality from mastitis are due to the breakdown of milk protein, called proteolysis. Cows with clinical or sub clinical mastitis show an increase in the activity of proteolytic enzymes that break down natural proteins causing bitter flavors, excess whey in yogurt and a decrease in cheese yields.

In organic production the optimum treatment of increased somatic cells is prevention. The remedy for this is simple — a clean dry environment. Housing of cattle needs to be kept clean, well bedded, with ample room for exercise and access to pasture in seasonal weather. Avoidance of wet, muddy, manure-dense conditions are critical to keeping a herd free from mastitis.

When a mastitis flare up occurs, those cows should be milked last to avoid transfer of bacteria to other cows, the milk should be isolated with a quarter milker, and the mastitic milk should not be fed to calves. Regular stripping of the milk from the udder can reduce inflammation and speed recovery.

The producer may use a strip cup at each and every milking as a quick screen for mastitis. This will flag any clinical cases of mastitis in the herd. Additionally an inexpensive test like the California Mastitis Test (CMT) helps the producer detect sub clinical cases and should be used regularly in addition to services such as DHI to monitor the herd. Regular culturing of individual animals with high somatic cell levels can help determine the source of bacterium. While the animal can combat strep-ag on her own, cows contaminated with *Staphylococcus* or "Staph" are generally culled from the herd, as they can not be treated with antibiotics and still be considered organic.

Antibiotics
Milk from a cow treated with any type of antibiotic cannot ever go into the bulk tank for organic production. However, animals must not be treated inhumanely, and if an antibiotic is essential for treatment, the animal will then be considered non-organic and most likely be sold. The milk from a treated cow cannot be fed to organic calves or other organic livestock.

Regular screening occurs for the beta-lactam group of antibiotics, which is licensed for use on dairy cattle. There are a wide number of other antibiotics which are not licensed for use on dairy animals that are often prescribed by veterinarians and sometimes neighbors--these include sulfa compounds, tetracyclines and gentamicins. These are all prohibited for use on organic farms. Many testing kits are becoming more sensitive for all types of antibiotics, not simply the beta-lactams, allowing the dairy plant to detect these prohibited substances. The FDA has announced that they are beginning to test milk nationally to determine the level of compliance to antibiotic use regulations. This should not be a concern for the organic producer, as these products are not allowed in organic production.

Preventative measures

Remember that an ounce of prevention is worth a pound of cure--a clean dry environment prevents the spread of bacteria to the cow. Not only should we address the environment within which the cow lives, but also the manner in which the cow is prepared for milking.

Wash water with the presence of a mild sanitizer is important in the preparation of the cow for milking. A cow that enters the milking parlor dripping wet from an underwater flush is teeming with bacteria. It is important that her teats and udder be washed AND dried prior to milking. In the stanchion barn, cows need to be adequately washed. It is important to use enough water to remove any dirt or manure from the teats AND the udder in proximity to the teats.

A critical step to the washing is the use of individual paper towels. While in some instances re-useable cloth towels that are washed between milkings can be used, it has been proven that the use of individual paper towels dramatically reduces the spread of mastitis-causing bacteria from one cow to the next. Environmentally friendly paper towels are available at most agricultural supply stores.

In addition to commercially available udder washes there are pre-dips and post dips available to the organic dairy producer. Those with serious mastitis challenges may greatly benefit from the use of pre-soaked individual wash towels as well as pre-dips and post dips. Others in different environments find that a good washing and a post dip is more then adequate to create clean milk. In either case it is important that the teats are clean and dry prior to placement of the milking machine.

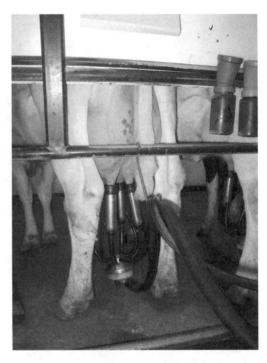

Clean udders to maintain milk quality.

> NOP 205.603(a)(4) Chlorohexidine is allowed for use as a teat dip when alternative germicidal agents and/or physical barriers have lost their effectiveness.

The milking barn and equipment

The most common point of contamination of milk with bacteria occurs in the process of milking, pumping and cooling the milk on the farm. The contamination is considered environmental, meaning that it comes from the air, soil, water, manure, bedding, silage, equipment and pipelines.

Parlors help maintain cleanliness during milking.

Foundations of equipment cleaning and sanitation

The basis for equipment cleaning and sanitation is to attack each of the essential elements that fuel bacterial growth. Bacteria growth, as we discussed earlier, is enabled by the presence of air, moisture, fuel (sugars, proteins or fat) and temperature. By countering each of these factors an effective sanitation program can be developed.

The relationship of time, temperature and concentration is important in adequate cleaning and sanitation. A decrease in any one of these three requires an increase in another component. For instance, in a situation where hot water is in limited supply one needs to increase the time and the length of the wash. There is a critical level where these adjustments are not adequate to effect competent cleaning.

The first step in cleaning equipment is to remove soil - this is done by an effective and forceful rinse cycle. Rinsing is important in that it removes heavy soils, which would dilute the concentration of the wash cycle to follow. This should be followed by a vigorous washing with a substantial level of caustic for the breakdown of fats and proteins, followed by a short rinse to avoid dilution of chemicals and then an acid wash. This should then be followed by a sanitizing step.

Rinsing

Rinsing should be done immediately after the parlor, pipeline or bulk tank is emptied. The purpose of rinsing is to remove free flowing milk solids and prevent any drying of solids on the equipment surface. Water temperature is not critical at this stage.

Washing

Washing should immediately follow the initial rinse. If it is necessary to delay the wash step, an increase in time or concentration may be needed to break down any biofilms (colonies of microbial growth) that may have developed. The purpose of the wash step is to breakdown fat, protein and milk sugars. It is important that the return temperature at the end of the cycle does not drop below 125°F.

Temperature Common wash temperatures in a milk line should be at least 140°F. A hot water set may be employed to boost hot water availability in the milk house. In a larger operation it may be necessary to increase water temperature above 155°F to ensure the complete cycle. Fat is water-soluble at this temperature and is easily removed as long as the wash water temperature does not drop below 125°F at the end of the cleaning cycle.

Time The force of the wash and length of the wash is as important as the temperature. Wash water is interrupted with bursts of air. This increases the turbidity or flow of the solution and also forces the solution to the top of the pipeline. With inadequate flow the inside top of the pipeline will not receive adequate cleaning. The length of the cycle varies with the size of the pipeline or parlor operation.

Concentration The third component to washing is concentration. Each chemical has a slightly different use requirement; the producer should read the label for proper usage. In an ideal situation a test kit will be used to measure the active ingredients of any cleaning solution in parts per milliliter concentration. This is VERY important, as more is not better. Too much soap is as problematic as too little. Most soaps are alkaline cleaners, with wetting agents that attack the protein and sugars in the milk. The alkaline also causes the fat component to become soap like while emulsifiers keep the fat in suspension.

Acid Wash

Depending on your chemical supplier's recommendation you may rinse between the caustic wash and the acid rinse. It is important to not mix acid and caustic washes or blend these chemicals. The blending of caustic and acid causes mustard gas, which was used in the trenches of World War I. Should an accident occur in the parlor or milk room, open all doors to the area and ventilate well prior to rinsing with copious amounts of water. The purpose of the acid wash is to remove the minerals deposited by milk and water. If a producer is using soft water, the acid wash may be a weekly occurrence. In most cases acid washes are performed after every regular wash.

Temperature Since the butterfat is removed in the washing cycle, the temperature of acid washes can be more moderate. It is still ideal to have return temperatures above 100°F.

Time Time and turbidity are as important in this step as it was in the wash step above.

Concentration Chemical manufacturer's instructions will dictate concentration. overuse of acid causes deterioration of gaskets and rubber parts in the milking system.

Sanitation

Sanitizing is the final step of the cleaning process. It is important that you only sanitize a clean surface. Sanitization of a dirty surface only masks a problem and is a poor use of resources.

There are a number of sanitation methods, some are approved for organic production; others are not.

Steam If you were in a plant environment with an enclosed system, steam would be the ideal method of sanitizing. By exposing stainless surfaces to steam or a temperature of at least 200°F for five minutes you have a natural organic sanitizing without chemicals. Unfortunately this is impractical at the farm level.

Chlorine Chlorine is perhaps the most widely used sanitizer available to the farmer today. It is effective on a wide variety of bacteria, works well against spore formers that are prolific in both the cold and warm bacterium classes, is not affected by hard water and is inexpensive. This form of sanitizing is optimum when performed just prior to use of the equipment.

Iodides Iodine is commonly used as an udder wash and a teat dip. Occasionally it may be used in washing systems. Some people with thyroid problems react badly to the presence of minute amounts of iodine, so this sanitizer should be limited in use in organic production. It does have an advantage over chlorine in that it remains effective on the contact surface for a greater period of time.

Acid Sanitizers Most of these sanitizers are phosphorus based. These sanitizers have two functions: they reduce the bacterial load, and prevent the build-up of minerals on the equipment. They have a 24-hour residual kill time, and when used with organic production the first milk through the line should be discarded. Because of the phosphorus base and the long residual impact these sanitizers should be reserved for particular regulated uses, not in a general use for equipment sanitation on a dairy farm.

Peroxyacetic Acid This is one acid sanitizer that I do like to recommend. This carefully developed blend of hydrogen peroxide and acetic acid (vinegar) creates one of the most natural sanitizers for use in the dairy industry. It functions well in hard water and is environmentally correct. As the sanitizer works through the system, oxygen is released with by-products of water and vinegar. It does have limitations against spore forming bacteria in certain settings, and should be balanced by an occasional sanitization with a chlorinated product.

Prohibited Chemical Sanitizers

Lastly in the "chemical" group of sanitizers is Quaternary Ammonia. This product is prohibited for use in organic production. While it is very effective against a wide range of bacterium, withstands heat and is most effective against thermoduric bacteria, it can cause off flavors in dairy products and seriously impede culture activities in yogurt and cheese. A common and appropriate use for this sanitizer is environmental — it is an effective sanitizer for the boots of visitors who may come to your farm and is commonly used in dairy plants to limit microbial traffic throughout the facility.

Hand Washing of Equipment

Hand cleaning is employed in most dairy operations — even if at minimal degrees. The cleaning and sanitation of quarter milkers, sick cow buckets, and miscellaneous equipment should not be ignored. In cases of erratic bacterial counts on a farm we often find that the use of a quarter milker or sick cow bucket is to blame.

Hand cleaning consists of the same basic principles as stated above: Rinsing, washing, rinsing and sanitizing. The selection of a good brush or scrub pad provides the force. Water temperature should be as hot as the individual can stand, preferably above 115°F. Gloves may protect the individual from the stress of heat on the hands. Buildups of milkstone can be addressed with mild dairy acids. On occasion a wash with a mild acid sanitizer also helps remove any buildups of minerals or milkstone.

Temperature and cooling

Milk as it exits the cow is approximately 100°F. This is a prime temperature for bacterial growth. Most bacteria, regardless of its preferences, will double within 20 minutes at this fertile temperature. One of most proactive things a dairy producer can do is to invest in a system that is adequate to rapidly cool the milk. The speed with which the milk drops below 38°F has a direct relationship to all types of bacteria counts.

In a small barn in-line cooling systems are often not practical. In these cases the bulk tank is the primary source of cooling. It is important that the tank be turned on shortly after the first cows are milked to begin the cooling. As this first milk cools, it naturally drops the temperature of the incoming milk. Agitation should begin after the milk reaches above the first two inches of the agitating paddle. If agitation is started too early, butterballs are created and an increase is seen in the acid degree value of the milk.

Proper maintenance of bulk tank is critical for temperature control.

Milking parlors and most larger barns holding over 80 cows generally have in-line plate coolers. These coolers are designed to drop the milk to less then 50°F prior to reaching the bulk tank. This increases effectiveness because the quantity of milk being placed in the tank would overload the ability of the equipment to cool the milk effectively. While the tank may reach the federal standards of less then 45°F in two hours, the microbes have had an ample start in the multiplication stage, causing the producer to produce sub par milk.

It is important to begin agitation of the milk at the onset of milking when a reserve of milk is in the tank. There are two reasons for this. First, it prevents stratification of the milk; the warm milk naturally rises — and is not cooled adequately when agitation is not immediately applied. The second reason is that you benefit from the transfer of temperature and reduce your cooling time and energy expended in the cooling process.

Understanding milk test results

Bacteria require 4 essential elements to grow:
1. Water — all bacteria require moisture.
2. Appropriate temperature — 5 ranges of temperature divides the types of bacteria.
3. Air or the lack of air — 4 gradations exist for the division of bacteria, relating to the amount of air required for growth.
4. Nutrition — sugars, fat and protein.

Milk is a good source of nutrients, has ideal moisture content and water activity. Bacteria, which attack milk products, can be characterized by which nutrient they destroy.
1. Proteolytic — Breaks down proteins
2. Lipolytic — Breaks down fat
3. Saccharolytic — Breaks down sugars

By-product of microbial activity are fermentation, gas production, stringiness, rancidity and abnormal flavors and colors.

Your lab results are a measure of bacteria in your milk. Careful review of your lab results can aid in the goal of producing clean milk.

Standard Plate Count

This is a measure of general bacteria in your milk. It is also called SPC or "Loop Count" and is the most basic test. It reveals general sanitation and health conditions on the farm.

Results:

Good	<3,000 per Milliliter (is reported XX per 1000)
Needs attention	<50,000 per Milliliter
Maximum Allowed	100,000 per Milliliter

Causes:
1. Failure to wash all milk contact surfaces after both night and morning milking with dairy cleaning compounds and HOT water.
2. Slow cooling of milk or failure of cooling systems
3. Poor sanitation practices in all milking and milk handling
4. Mastitis in the herd.

Preliminary Incubation Test

This is a measure of the bacteria that will reproduce at refrigeration temperatures. It is also known as a "PI" or Psychrophile test. It is one of the most important and critical causes of poor shelf life in fluid milk.

Results:

Good	< 1,000 per Milliliter (is reported XX per 1000)
Needs attention	<5,000 per Milliliter
Maximum Allowed	50,000 per Milliliter

Causes:

1. Failure to clean milk contact surfaces properly
2. Old, cracked inflations and rubber parts, milk pump seals bad, line gaskets in place too long.
3. Bulk tank not cleaned properly
4. All milk contact surfaces must be clean and sanitized
5. At milking, udders and teats must be clean and dry.

Pasteurized Count

Also known as LPC, this is a count of bacteria types that survive pasteurization under laboratory conditions. These bacteria reduce shelf life of packaged milk, cause early spoilage and off flavors — they lose customers.

Results:

Good	< 50 per ml
Needs attention	<100 per ml
Maximum Allowed	250 per ml

Causes:

1. Wash water not hot enough
2. The use of non-dairy cleaners for washing
3. Poorly cleaned bulk tank (milkstone or water stone buildups)
4. Milk lines with a film inside
5. Vacuum lines and air hoses with buildup inside
6. Old, cracked or corroded rubber parts
7. Use of *Bacillus* bacterium as a natural parasiticide.

Coliform Count

This test is also called coli count. It is a test for bacteria indicating extreme unsanitary conditions. Coliform bacteria in milk are an indication of contamination from animal waste products.

Results:

Good	10
Needs Attention	25
Maximum Allowed	100

Causes:
1. Drippings from washed but undried udders
2. Inflations not changed regularly
3. Deterioration of inflations from inadequate or extreme chemical concentrations
4. Contaminated water supply
5. Milk stone in tank or milk lines
6. Failure to replace gaskets in lines often enough.

Additional Lab Tests
There are a few other tests which are used frequently for the analysis of the quality of the milk supply. The purpose of these tests is not microbial but is compositional in nature.

Sediment
This test reveals foreign materials in the milk. A very minute amount of foreign materials are allowed in milk. So much as a single fly is considered a level 3. FDA & USDA both require farms to be tested for sediment on a regular basis. Processors may require this test to be performed daily or weekly on routes that are problematic.

This test is performed by screening the entire bulk tank of milk through a plastic or fibrous screen. It is evaluated on a scale of 1-4.
Results:
    Good   1
    Needs Attention   2
    Maximum Allowed   3

Causes:
1. Unwashed udders
2. Unclipped udders, seasonally shedding cows
3. Poor seals on bulk tank
4. Poor seals on doors or windows of milk room
5. Allowing milking units to touch or fall on the floor or into bedding.

Acid Degree Value
This rarely used test reveals the breakdown of butterfat due to excessive agitation or pumping of the milk.
Results:
    Good   0.90
    Needs Attention   1.10
    Maximum Allowed1.50

Causes:
1. Anything which may cause over agitation of milk
2. Agitator left on after milk has cooled.
3. Improper seals in milk lines allowing air bubbles or leaks
4. Malfunctioning milk pump incorporating air with milk
5. High milk line requiring excessive lift of the milk
6. Large number of cows in late lactation in the herd
7. Freezing of milk in the bulk tank.

Cryoscope Test or Freezing Point

This test detects abnormal amounts of water in milk. This test is performed by quickly freezing a small vial of milk. This is the most accurate method of measuring percent of water in milk. Normal milk will vary slightly in freezing point but it is considered abnormal or contaminated as the freezing point becomes closer to 0°C.

Results:

Good	> - 0.540
Needs Attention	> - 0.535
Maximum Allowed	> - 0.525

Causes.
1. Pipelines not sloped correctly, allowing for pooling of milk and rinse water
2. Water left in equipment or rinse water allowed to get into milk supply
3. Careless washing of the top of the tank with a water hose
4. Serious mastitis problems can alter freezing point
5. Intentional

Organic dairy production is about doing things right. From fresh feed to clean barns and roomy pastures, good milking practices and an effective cleaning and sanitizing protocol creates a premium organic dairy product. The best milk is that with the least contaminants. As organic dairy producers we pledge to keep out the world's hazards, from GMOs to pesticides, synthetic fertilizers, hormones and antibiotics. We also pledge to produce milk as natural and clean as nature intended. Paying attention to details of quality can lead to success.

Farmer Profile: Jim Greenberg - Stratford, WI
Jody Padgham

Jim Greenberg of Stratford, WI has proven he's a man who isn't afraid to try new things. He was raised on the farm in north central Wisconsin that he still operates with his daughter, two sons and their families; but his ancestors would never recognize the place today.

"We've gone through a lot of changes over the years, and are very happy with where we are now, grazing and certified organic," Jim's son Jeff Greenberg, 34, says. Five families now work and live off the income from the farm.

The story of how Jim got into large-scale organic grass dairy is an interesting one. In the 1990's Jim left the farm, went east and landed on a very large free-stall dairy in Michigan. He returned home after several years, with "bigger is better" on his mind. Over the next several years Jim built large holding facilities and a double 20 milking parlor so that he could grow the family farm. "In the 1990's we had plans for a 1000 cow free-stall set-up," Jeff tells me. "We put the cows outside in the summer — 300 hundred cows got 15 acres!" Cows were milked three times a day; a factory-like efficiency prevailed.

Jim soon realized that he wasn't happy with the dairy system he had set up. A reader and lifelong learner, his research pointed him toward the cost efficiencies and improved herd health gained through grazing. "Dad always wanted to pasture," Jeff says. "We never were big into chemicals, and so once we took the step to grazing, organic was a logical next step." Now the cows are outside year round, rotationally grazed on paddocks, and the free-stall barns have been reclaimed for other uses.

"It's harder coming from so big an operation with huge overhead into organic grazing," Jim notes. "We have more infrastructure than we need, and are still paying for it, but the organic premium helps. We are really happy to be where we are today — this is the right way to farm."

The Greenbergs manage around 2000 acres, with about 640 acres in pasture and the rest in cropland. "We are one of the few dairies you'll hear about that are trying to downsize," Jim tells me. "400 cows would be ideal for our situation. We are a little tight on pasture at each end of the season, and are still cutting back on the herd."

Jim Greenberg hosts a field day.

Grazing paddocks are 10 acres each, and they rotate two groups of milking cows every 12 hours on a 10-day rotation. Paddocks are set up permanently with watering fountains in each. "I might have done it differently if I was to do it over again," Jim says. "The permanent water is easy, but dictates the paddock — I would like more flexibility." Covering the 640 acres of grazing land is no small feat. During peak season one person spends 8 hours a day just moving cows. "We put 2000 hours a year on the Gator 4-wheeler just moving cows," Jim says.

The Greenbergs found that it took the cows a few years to adjust to the pasture environment and a few years for the pasture to adjust to the cows. "Cows that have been inside basically all their lives need to get used to being out in the sun and heat," Jim says. "The cows have slimmed down since they are out on grass. They look better, sleeker."

Cows are run out in paddocks through mid-November. When the grass really stops growing, they are put into 5 groups and allowed to roam on a manure pack alongside one of the old free-stall barns. "I sure can see the difference between having them out on this pack and the free-stall we used to have," Jeff notes. "There would always be a few that would get caught up in the stalls every day. They really seem to like to sprawl and choose their position." The Greenbergs pay more in the long run for bedding now, having to purchase sawdust for the pack but believe the increase in cow comfort is a good tradeoff.

Calves are raised up on the farm, with over 400 born each year. The Greenbergs have not bought a replacement in the years they've been grazing, and in fact find a good market for any extra heifer calves they have. "We have cows on the farm that we've been milking since 1994," Jeff notes. "We used to push them more, now we take it easier." Herd average right now, without any protein supplements, is down to about 40 pounds of milk per cow. It used to be 65 per cow. "We're happy not pushing the cows, extending their life."

Cow health has dramatically improved since the free-stall days. Jeff notes that the sun and grass just make for healthy cows. "Those cows that were inside 24/7 were prone to problems. The difference is remarkable." The Greenbergs work with homeopathics and other formulas on the rare occasion they see a health issue.

The Greenbergs went through an 80/20 herd conversion to organic in 2003-2004 and started shipping certified organic milk in early 2004. "We were using so few chemicals, and the organic premium was attractive, so we decided it was the right thing to do," Jeff says. "We're into organic 100 percent — Dad tries to buy all organic at the grocery store and I'm always reading labels. Organic makes sense."

They shopped around for a milk shipper once they decided to go through the transition, and choose to work with Horizon Organic Dairy. "Horizon has been very good to work with," Jeff notes. "They communicate with us frequently, asking how things are going." Jeff tells a story about their first few months of shipping organic milk. "We had put in a new bulk tank and were having some problems getting the adjustments right. Horizon was very patient and helpful as they worked with us while we figured out the problem." Jeff comments that the increased number of organic milk marketing groups is good for the farmers. "It used to be folks would be on a long waiting list. You don't see that so much anymore." Along with that, Jeff mentions that Jim has been attending meetings regarding a Midwest version of the Northeast Organic Dairy Producers Association, a group that communicates about price and supplies of organic milk. Jeff believes that organizations like NODPA are important so that "we don't see oversupply and price deflation like we've seen in conventional milk and other commodities."

This farm tour would not be complete without talking about the dairy parlor. Installed in 1998 in one of the old free stall barns, Jeff says that the DeLaval double 20 parallel is "comfortable, but not fancy." Two people milk about 120 cows an hour. One thing that does seem fancy, however, is the computer system that runs with the parlor. Each cow carries a computer chip that is read as the cows move through the parlor. Production is automatically tracked. As the cow leaves the parlor, any animal with problems or health issues can be automatically funneled to a holding chute in response to a trigger from the computer.

It is hard to capture in words the many innovations and inspiring things that are happening at Greenberg Farms in Stratford. This extended farm family is a great example of how those not afraid of change and willing to move in a new direction that may not follow that of their friends and neighbors can really succeed. The pressure of debt and infrastructure keeps their decision-making tight and their pastures a little overfull, but the Greenbergs know that they have made the right decisions in managing their large herd of cows organically and on pasture.

Chapter 7

Farm Biosecurity

Tamiko Thomas

In recent years major infectious disease outbreaks both here and abroad, combined with increasing anxiety over bioterrorism, have made biosecurity issues on farms a growing concern. Diseases such as contagious mastitis, bovine viral diarrhea, Salmonella infection, and Johne's disease result in production losses, premature culling and a potentially significant impact on the bottom line.

It makes good sense for dairy farmers to have effective biosecurity procedures as a preventative measure. In particular for organic dairy farmers with their reduced arsenal of weapons against disease, prevention is the key. It is stressed in the NOP Final Rule that producers must establish and maintain preventive livestock health care practices. This will help protect the health and welfare of their cows and secure the future of the farm.

Developing a plan

An important key to developing an effective biosecurity program is assessing the risk of disease entering the herd and forming a prevention plan with input from your herd veterinarian. The plan should be discussed with your family and any employees. It is important that they understand the why and how of a biosecurity plan for it to be effectively implemented. Regular visitors should also be informed of the precautions. Risk factors to be considered in such a plan include importation of new animals, movement of people and vehicles and contact with other animals.

Importing new dairy animals

Bringing in new cattle has a high potential to introduce new pathogens onto the farm. Maintaining a closed herd is ideal. However, if this is not possible then sourcing cattle from a herd with a known health status is important. Dairy cattle should be tested and vaccinated for relevant diseases prior to being brought onto the farm. Testing of individual cow milk samples and/or bulk tank milk samples is recommended to prevent the introduction of contagious mastitis organisms. In order to maintain this known health status it is important to eliminate contact during transit from the farm of origin with other dairy animals and their manure.

Quarantine of new, or returning, dairy cattle is one management practice that can reduce the likelihood of introducing certain diseases to the herd. New arrivals should be quarantined so that they don't come into contact or share air space or equipment (feeders, waterers, etc.) with your on-farm cattle. Further precautions should be taken to ensure that pathogens are not carried on coveralls, boots, hands, etc. between the quarantine cattle and resident cattle. A suggested quarantine time period is 21 to 30 days. Quarantine should be used in conjunction with other biosecurity measures because some diseases such as Johne's are difficult to find early on and clinical signs do not usually appear until 2 to 5 years after exposure.

Another important component of a solid biosecurity plan is to increase the current dairy herd's resistance to disease. This can be done through breeding programs, vaccinations and colostrum management. In addition, reducing potential stressors by providing proper nutrition, proper handling, a suitable environment and removing internal and external parasites is crucial. Stress can pull the animal's immune system down and increase the animal's susceptibility to disease.

People and vehicles

Disease can also be brought onto the farm by people and vehicles. The primary objective should be to restrict access to the barn and other cattle areas. Clearly visible signs should be placed at the dairy entrance directing all visitors to a specific contact point on the farm before they go into the interior of the dairy. Keep any dealers or transporters out of your barns when they are picking up cull cows or bull calves. Vehicles and equipment that your animals are to come in contact with should be sanitized before coming onto the farm. Move dead animals to a specific location for pick up thereby keeping the livestock renderer away from your barn and from coming in contact with your animals.

Plastic boots for visitors help prevent farm-to-farm contamination.

It is imperative that if visitors are to enter the barn you make sure they have not had recent contact with farm animals on other livestock facilities. Ask them if they have recently done any international travel. International travel is an important consideration because foot and mouth disease is endemic to a number of countries and can easily be spread. People who have traveled to countries where foot and mouth is present should be prevented from entering the farm for at least 5 days after their arrival in the United States. If they are allowed to visit the cattle they should wear protective clothing and dairy foot gear or their own foot gear should be thoroughly disinfected prior to entering the operation. Visitors should be prevented from having close contact with or handling animals.

Other animals

Contact between dairy cows and other animal species, both wild and domestic, should be limited. In particular it is important to restrict the access of such animals to stored feed and feeding areas. For example, it is important to keep dogs out of the dairy feed because they can spread the protozoa *Neospora caninum* via their fecal matter. Infection can cause abortions in cattle.[14]

Inadequate biosecurity allows disease to come onto the farm. Diseases carry a high cost for the producer and the animals and therefore proper biosecurity measures are a must.

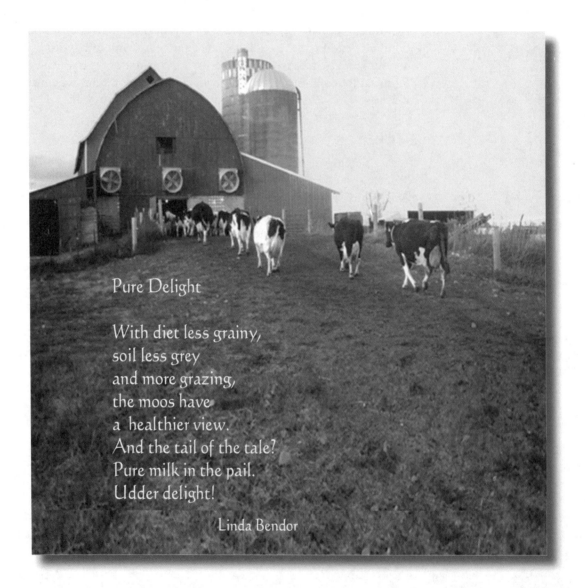

Pure Delight

With diet less grainy,
soil less grey
and more grazing,
the moos have
a healthier view.
And the tail of the tale?
Pure milk in the pail.
Udder delight!

Linda Bendor

---

[14] Kirk, J.H. *Infectious Abortions in Dairy Cows.* UC Davis Website: www.vetmed.ucdavis.edu

Chapter 8

Animal Breeding for Organic Production

Robert Hadad

Breeding for organic production is not a new concept--for some time now a good deal of thought has been given to breeding plant varieties, particularly vegetables, that are more suitable for organic production. With an eye towards greater sustainability, relying on genetic traits that work with nature makes a great deal of sense.

Much of the readily available breeding stock comes from the conventional side of agricultural production, where high yield is the driving force for selection. This is particularly true with dairy cattle. The traits important for organic farming (such as health, fertility, mastitis resistance, longevity, forage intake capacity, feet and leg strength, lameness resistance, parasite resistance, and hardiness) are generally deemed of lesser importance in conventional breeding stock. However, in organic systems these "lesser" traits can be as important as production.

In conventional agriculture, moving towards standardization has been the norm for crop as well as animal breeding. Fewer and fewer varietal differences are being introduced and biodiversity has been reduced. This goes against the principles of organic agriculture, which encourages biodiversity and uses biodiverse selection to adapt to the local ecology. According to the National Animal Health Monitoring system of the USDA, 93.4% of dairy cows in the U.S. are Holsteins. 80 % of these Holstein cows are bred through artificial insemination (AI) to only about 20 sires or their sons.[15] (A summary of alternative cattle breeds appropriate for organic dairy operations can be found in Appendix B.)

---

[15] *Dairy 2002 Part II: Changes in the United States Dairy Industry, 1991-2002.* USDA Animal and Plant Health Inspection Service, Veterinary Services, National Animal Health Monitoring System, June 2003.

Genotype-environment interactions

The same way that heritage or heirloom vegetable and fruit varieties have improved opportunities for farmers in the marketplace and by matching crops that are more adapted to the local environment, local, heritage, or minor breeds of livestock can fit into an organic farming system. The original development of heritage livestock breeds was in response to selection by environmental conditions. These conditions might have been climate, localized weather, vegetation type, vegetation availability or seasonality, soil conditions that affected the nutritional makeup of forages, predator pressure or type of use by the farmer selecting animals to keep. For example, within a particular genotype, milk production based on a diet exclusively of forages will not be the same as milk production on a diet comprised predominantly of concentrates. Since cows have been kept mostly confined and generally separated from the local environment, genotype-environment interaction traits have been greatly reduced or even lost in recent times.

Breeding strategies

Two types of breeding strategies available to organic farmers are "on-farm breeding" selection and quantitative selection. On-farm breeding selects animals based on their response to the conditions on an individual farm. This type of selection is a long process but evaluates animals under the rigors of the local environment and gives site specific results. Breeding of animals using quantitative genetics produces an average animal based on the population from which the parental selection lines come from. This approach takes a shorter amount of time but the value obtained for the individual farmer in a given location will be less than site specific selection. Other breeding schemes combine these approaches. One combination is called foundation breeding.

Dutch Belt, milking shorthorns and Jersey cross are part of this organic farmer's herd.

Foundation breeding uses purebred animals and follows the on-farm breeding scheme.[16] Groups of farmers work together with each farmer participating creating a unique population within the total population of purebred cattle. Then farmers choose bulls from each other's herds and create their own crosses with different population lines.

In Holland, organic dairy farmers are using foundation breeding.[17] One group of farmers are breeding purebred Dutch Friesian cows. The cows on these farms

---

[16] Nauta, W. 2001. *Breeding Strategies for Organic Animal Production – an International Discussion.* In *The Proceedings of the 4th Network for Animal Health and Welfare in Organic Agriculture (NAHWOA).* Wageningen, The Netherlands.

[17] Endendijk, D., Baars, T. 2001. *Family Breeding at Rivelinohoeve.* In *The Proceedings of the 4th Network for Animal Health and Welfare in Organic Agriculture (NAHWOA).* Wageningen, The Netherlands.

become more and more adapted to their own specific farm's environmental conditions as well as the breeding goal of each individual farmer. Due to this, variations between farms develop within the breed. Bulls from other farms can be used in rotational crosses. As the need arises, different breeds might be added to the breeding scheme to improve traits that might be deficient in the purebred lines.

Breeding values

With conventional dairy breeding, performance traits such as milk yields and composition predominate, with only minimal attention paid to functional traits. Functional traits must carry more weight in an organic dairy breeding program. Traits such as longevity, mastitis resistance, and persistence of lactation are quite important. Other traits such as health, leg conformation and resistance to lameness, conformation of udders, milk quantity, milk fat, milk protein and increase in milk production from the first to the third lactation are also very important. Also key are the rate of living calves at birth and the amount of time from calving to the next successful breeding (service period). With organic dairy production relying on pasture feeding, finding animals that perform well on a grass-based diet is essential.

The Swiss, who have done a lot of work in genetics for organic dairy, have identified four main issues that are keys to developing a successful organic breeding system:

- Breeding to meet the needs of organic dairy farmers is a long-term process. An organized effort of research and sharing of experiences among farmers is key.
- The breeding scheme must focus on functional traits rather than only looking at high yields. Getting all the dairy breeders on board with this goal is essential.
- Organic dairies should purchase stock from other organic dairies that are following the same principles while following good biosecurity protocols.
- Organic dairies should remain true to the goals of an ecological organic breeding program and not follow the latest selection trends.

Mastitis resistance

Selection for mastitis resistance has definite benefits for an organic dairy breeding scheme. Production practices and sanitation can go far in the prevention of mastitis but also having cows that have genetic resistance will ensure both economic and welfare benefits[18]. It has been found that selection for high yield alone can lead to substantial reduction of functional traits, including mastitis resistance.

---

[18] *Christensen, L.G. 1998. Possibilities for Genetic Improvement of Disease Resistance, Functional Traits, and Animal Welfare. In: Breeding for Disease Resistance in Farm Animals. CABI Publishing 2000.*

Keeping cows free from disease and other problems that influence well-being is cost effective, especially when the genetic traits of the animal are responsible for the defense rather than purchased inputs. Despite the relatively low heritability, with time, major economic progress can be made by selecting for mastitis resistance.

Parasite resistance

One area where breeding for functional traits would be very beneficial is for parasite resistance. Ruminants tend to be susceptible to a variety of parasites, and it is hard to keep livestock parasite-free, particularly in a pasture-based system. Chemical de-wormers are commonly used on conventional dairy farms but are not allowed in organic systems. Organic producers have tried herbal remedies or diatomaceous earth with some success. Research for approved parasite controls is a high priority for organic livestock producers.

Unfortunately, there is not a great deal of research available on progress towards parasite resistance through breeding in dairy animals. However, similar attempts have been undertaken with sheep in Australia.[19] A current project has provided some early evidence that breeding for resistance is possible, although it is a long process.

Selecting for resistance to a particular species of parasitic worm in sheep has also yielded the benefit of providing resistance to other parasitic worm species. This broad resistance is termed cross resistance. Cross resistance is important because parasite exposure by number and species changes from year to year.

Creating a breeding program that includes parasite resistance would have substantial benefits for an organic dairy operation. Through the use of resistant animals, fewer parasites would complete their life cycles, resulting in decreased parasite populations in the pasture. With reduced parasite numbers, it would be harder for susceptible animals to become infected. With a closed herd, new introductions of parasites would drop dramatically.

Crossbreeding

Although there is a great deal to be said for maintaining purebred lines, especially for heritage or rare minor breeds, crossbreeding for the production herd allows for quicker genetic improvements. Increased vigor is often seen from the crossbreeding of two differing parental lines. Introducing a trait from one line into another where that trait is missing is the goal. One example is for polled cattle. A single gene controls polledness which does not normally occur in the Holstein breed: however, it does in several other breeds, such as Dutch Friesian or Red Poll cattle. Through crossbreeding, the polled gene can be introduced into a normally horned breed creating a new polled crossbreed. This means that calves do not have to be dehorned or debudded. This procedure would reduce stress for the calves and save the farmer time and money.

---

[19] *Eady, S. 2002. Evidence to Support Breeding For Resistance is an Effective Component of Sustainable Worm Control. Commonwealth Scientific and Industrial Research Organization (CSIRO). www.csiro.au*

By working closely with your herd, and over time understanding the traits that work particularly well for your farm and environment, you can improve herd health and production through breeding. Consider one or several of the methods of breed selection listed above, and find the path to genetic specialization that works for your place and situation.

Farmer Profile: Cheyenne Christianson - Chetek, WI
Jody Padgham

Cheyenne Christianson farms with his wife and six young children near Chetek, WI. Cheyenne manages 240 tillable acres on his farm, rents 25 acres and milks 60 Holsteins. He has been grazing since 1993 and has been certified organic since 1999. He also has the unique claim of feeding no grain to his cows since 2000. He astounds many by maintaining high herd health and a herd average of 12,000 pounds per year. His farm is very profitable, with no vet on the place for three years, few health issues and very low feed costs. Cheyenne raises most of his own forage, but has brought in a semi load of quality western hay as a supplement each of the past few years.

When speaking of genetics, Cheyenne likens his cow herd to open pollinated corn. He has switched in the last several years to using his own bulls, to capture the genetic specialization he sees in his herd. "Using artificial insemination (AI) is like running your fertility program off of NPK," Cheyenne says. "AI will give you the generic best — but I have a very unique system here, and I want cows that work well for my system."

Using stock bred from his non-grain herd that is outside year round in his cold climate, he finds that the calves, though smaller and slow growing, are full of vigor and continue growing long into their life. His goal is to breed for 1200 pound Holsteins that will milk well with no grain. "A smaller cow eats less, but won't necessarily produce less," he notes. His Holsteins maintain an average of 4.1% butterfat. "My next goal is to move into line breeding," he says. "There is a concern with inbreeding, so this will be the way to go."

Fighting personal health challenges the past few years, Cheyenne has been more or less forced to slow down the intensity of his management. "The funny thing," he says, "is that the simpler my system gets, the more herd health improves and the more profitable the farm gets." Cheyenne believes strongly in feeding his minerals by feeding the soil and letting the plants bring the minerals to the cows. "Cows can utilize most of the minerals a plant grows, while they can only use a portion of a supplement."

He does, however, feed an organic approved free choice mineral mix, though he says his cows don't eat much of it. He is a strong advocate of kelp, high in minerals and iodine. Cheyenne fed out three tons of free choice kelp last year. He manages his pastures by soil testing and applying trace minerals from Midwestern Bio-Ag, using hi-cal lime and spreading basalt rock screenings at a rate of one ton per acre. He moves cows from the manure pack to pasture as soon as things green up, continuing to feed round bales on the pack at night for a while. Bedding is old hay. Cheyenne is a strong believer in mature pasture, and likes a stand to be at least two feet high before the cows get to it.

With such a low cost of feed and organic premiums for his milk, Cheyenne finds that he has very high profitability on cows with rather moderate yield. "I have the newest equipment of any farmer I know" he says. "With a steady income, I need to spend money, and I don't like to fix things." We all should have such problems!

Chapter 9

A Biological Approach to Soil Health

We now move from the dairy animal to the rest of the farm system. As we noted in the introduction, a successful organic farmer should spend as much time managing soil and crop systems as in the barn with the cows. A healthy organic dairy system is built on the foundation of quality feeds, which must come from well managed, balanced soils.

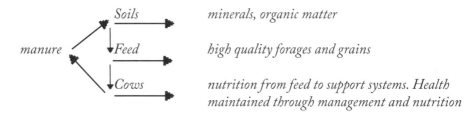

In this chapter we will give you an overview of the concepts of biological soil balancing--feeding soil organisms so that quality crops will be produced. We will detail the basics of taking and interpreting soil tests, which will help you gauge the quality and balance of your soils. We will conclude this section with a summary of the importance of soil organic matter.

Biological Soil Balancing
Gary Zimmer

Healthy soils produce healthy crops which, when fed to livestock, keep them healthy and productive. The objective of biological farming is to have soils that are rich in a balance of minerals, which are then available to the plants that grow on them. Biological/successful organic farming is a totally different "system" of farming, using different rules than conventional agriculture. It requires diverse knowledge, common sense, new thinking and intense management.

Biological farming is farming that utilizes both science and nature to create superior agricultural systems. It improves the environment, reduces erosion, reduces disease and insect pressure and alters weed pressure while working in harmony with nature.

When using the word "organic" to describe a production system, the farming program is based on the use of crop rotation, crop residues, compost, and organic materials to maintain fertility. To also be sustainable means you must maintain some level of soil health, quality and mineral level. If the farming method depletes these resources, soil is lost, and damage to the water supply can also occur. The National Organic Program Subpart C- Organic Production and Handling Requirements 205.200 General states: "Production practices implemented in this subpart must maintain or improve the natural resources of the operation, including soil and water quality." Being sustainable and protecting the natural resources (like minerals) are not only required for organic farmers but are essential for success. For an organic farmer, there is no choice but to maintain soil quality. A farmer should not need a rule to force him or her to take care of the soil; it should be a primary objective in their farm operation.

Healthy soils produce healthy crops.

Minerals, sunshine, water and soil life are the things needed to grow good crops. Biological farming is basically about soil life and healthy soils. Soil quality, and soil structure are all related to soil life. Soil organisms, such as earthworms, nematodes, bacteria, fungi and other microflora are the bridge between organic materials, soil minerals and healthy crop production. Very little is known about the total working of the soil, what lives there and what these organisms do. What we do know is that you, the farmer, can influence not only which organisms, but how many of these vital soil inhabitants live and work for you in your soils.

This section is divided into four management areas.
1. Minerals: testing, balancing, mineral use, manure use, soil corrections and crop fertilizers
2. Rotations: crops, their rotations with fertilizer and manure management
3. Tillage: for controlling air and water, the decay of residues and weeds
4. Feeding: feeding and taking care of soil life

Mineral management
Every soil is different, because it was formed with different minerals. All minerals are needed for crop growth and health, and these same minerals are

needed by the dairy animal. One method of assessing soil fertility is the success of livestock production without supplementation. What's in the soil will affect what is available in the feed grown on that soil.

There are various methods of soil testing and a variety of things you can test for. Most testing labs will come up with consistent conclusions that specific nutrients are short or in excess in your soil. Different labs will, however, often show different actual amounts of each nutrient. Soil tests don't actually measure what's in a soil, they measure what can be extracted by a test in a lab using a mild acid. The lab technicians then estimate what a crop would be able to get from this soil. This is why over a span of years it is a good idea to compare tests from the same lab.

Unfortunately, soils don't fit into a perfect mathematical formula. If your soil test shows X number of pounds short of a mineral, you can't just assume you need to apply X amount and all will be well. How much and how fast to apply a nutrient depends on many factors, money being one. If the soil test says you need nutrients, then supplying them is money well spent; you are heading in the direction of soil balance. The type of soil, the amount of the nutrient used, the crops grown, and the nutrients removed will all influence the speed of soil corrections.

Though there are many different minerals needed by soils and plants, most labs will test for only ten of these, which fit into three groups: major, secondary and trace elements. All of these are equally important; what separates them is the amount of each that is needed. The ten commonly tested nutrients are phosphorus, potassium, magnesium, calcium, sulfur, zinc, manganese, iron, copper and boron. We can test for other nutrients, along with other useful soil information like cation exchange (nutrient holding) capacity (CEC), organic matter, pH, base saturation and ratios.

Because there are many more minerals needed for healthy crop production than just the ten we test for, using natural mineral sources like rock phosphates and kelp when correcting soils can provide additional vital nutrients. Kelp, for example, contains 70 vitamins and minerals, such as selenium, which is needed in very small amounts.

Dividing mineral management into two areas, soil correction and crop fertilizers, makes the job easier and gives great results.

Soil Correction
The first area of mineral management is soil correction. Based on your soil test, whichever minerals are short should be added, though generally not all at once (unless you have lots of money). An organic farmer can use any naturally mined mineral any other farmer does; he or she just needs to test to make sure it is needed. Minerals must, however, come from natural sources that are slow release and depend on soil life and/or green manures to do the breakdown work.

I recommend you start your mineral management by looking at calcium and phosphorus. Calcium is our primary "tool" in biological farming. Calcium affects soil structure, soil biology, mineral uptake, plant health, plant energy and livestock health. From your soil test you can choose the correct sources for your situation:
1. If you need both phosphorus and calcium start with rock phosphate. It supplies both.
2. If the pH is low and calcium is low, "lime" your soil, taking care to choose the correct type of lime and the amount needed for your soils. Certified organic systems can not use hydrated lime, but can use mined lime.
3. If the pH is 7.5 or higher, calcium is low and magnesium high, use gypsum. Again, mined gypsum must be used in certified systems; recycled wallboard is not allowed.

Crop Fertilizer
The second area of mineral management is crop fertilizers which provide nutrients to the crop at levels above and beyond what the soil can provide. Placing these as blends in the plant root zone area gives you the greatest effect.

Each crop has specific nutritional needs which must be met. Fertilizers to address these needs won't necessarily correct the soils (change them over time) but will provide the needed crop nutrients until the soil is completely balanced. A crop fertilizer is a balance of all nutrients specifically formulated for that crop. Some minerals, like nitrogen (the only mineral you can grow), sulfur and boron, are always needed because they are anions and leach out of the soil. Twenty-five pounds of sulfur and one pound of actual boron is required per acre for all crops each year. You are never done with crop fertilizers, because you harvest and sell nutrients down the road in the form of milk, meat and crops. Other materials leach away, and still others, because of grazing and cattle management, get placed via manure in areas where they are not needed. Buying a lot of feed brings in nutrients; selling feeds, especially forages, removes not only organic matter but also many minerals from your system.

Rotations

Dairy cows are designed to eat forages. Organic dairy farmers use a lot of forages. A large part of a biological dairy system should be in rotationally grazed pastures and forage crops. Organic farmers can't easily go out and buy supplements for success; they need to grow super quality forages. If the goal is to feed 75 percent of the diet in forages, then I recommend you plan that 75 percent of the land be in forage crops.

A successful rotation will start with a seeding year, either with a small grain crop (such as oats) or forage crops such as triticale, oats, barley and peas. Two years of forages are followed by a year of corn. This tight rotation helps with soil mineral balance, and allows you to grow great corn crops easily. Next, working down a beautiful alfalfa, clover, and/or grass forage with manure on it not only provides a lot of nutrients for the next year's crop, but with only one year of corn, weed control is easy. There are no fresh weed seeds, and the forage absorbs the soluble nutrients from manure so they can't fertilize the weeds. Three years of forage crops means undisturbed soil, which are ideal for soil life, creating loose, crumbly ground with reduced weed seed germination and growth.

Crop rotations build soil.

What's above ground affects what's below ground; certain plants feed certain populations of organisms. An important key is crop diversity, not only for insect and disease control, but for mineral exchange and healthy soils. Quality grasses and other legumes besides alfalfa are now available. Typical mixtures might include rye grass, festulolium, soft fescue, some orchardgrass, alfalfa, red clover and some white clover.

Forage crops remove the nutrients (and organic matter) from the land. Raw manures re-introduce soluble nutrients, especially nitrogen and potassium. This can lead to forage imbalances, insect problems and weeds if not managed properly. If you must use manure on hay fields, do it after the last cutting or grazing in the fall. Composted manure, because of the stabilized nutrients, can be supplied any time to hay fields with beneficial effects. Organic matter bedding and manure pack composting fit any farm, but especially an organic farm. It's a good place to grow beneficial organisms and help maintain and supply minerals to the crops. Composting speeds up the decay cycle in a controlled environment and reduces the volume of materials needing to be hauled.

Keep in mind that the National Organic Program (NOP) has a very strict definition for the substance called "compost" versus manure, and has very strong restrictions for the use of manure on any crops that are consumed by humans. Crops grown for animal feed have few restrictions on the use of manure.

Tillage

Farmers have soil jobs to do. They must control air and water and manage the decay of residues. Soil life "wants its food on top and wants to be left alone." Over-aeration of soil burns up organic matter and damages soil structure. Luckily, forage crops don't generally have weed problems and don't require major soil disturbance. Organic dairy farmers generally till only to kill the forage plants for rotation and for weed control.

Shallow incorporation of residues makes sense. Fortunately, we have tools today that can do the job with minimal impact on soil life. Many organic farmers have success with tools like rotavators. They allow less tillage and the ability to aggressively kill forages (alfalfa, clovers, grasses) in one pass with just surface tillage. (For more information on cultivation equipment, refer to

Composting of Manure

The NOP has very specific definitions for any compost that will be used on crops consumed by humans. Since most dairy farmers will use compost as an enrichment for crops that will be fed to cows, they do not need to follow the NOP rules for the production of compost, which are generally considered quite stringent for the production of quality compost. HOWEVER, if you do not follow the NOP recommendations for compost production, the manure, if composted or not, must be applied according to the rules stated for "raw manure." This should not be a problem for organic dairy systems.

The NOP definition of compost is as follows: §205.2 "The product of a managed process through which microorganisms break down plant and animal materials into more available forms suitable for application to the soil. Compost must be produced through a process that combines plant and animal materials with an initial C:N ration of between 25:1 and 40:1. Producers using an in-vessel or static aerated pile system must maintain the composting materials at a temperature between 131°F and 170°F for 3 days. Producers using a windrow system must maintain the composting materials at a temperature between 131°F and 170°F for 15 days, during which time, the materials must be turned a minimum of five times."

Stockpiling manure is not adequate to meet the compost regulation. Manure is not "compost" unless it meets this definition. But, since crops grown on dairy farms are generally fed to livestock and not for human food, this should not be a problem for organic dairy systems.

For more information on composting, we recommend the publication "The Art and Science of Composting: A resource for farmers and compost producers" by Leslie Cooperband, University of Wisconsin-Madison Center for Integrated Agriculture Systems, 2002. It may be obtained on the CIAS website at www.wisc.edu/cias/ or by calling 608-262-5200.

the book *Steel in the Field* by Greg Bowman, published by the Sustainable Agriculture Network.) Remember, however, that there is no one way to farm nor one tool to farm with. If major soil corrections are needed and/or a lot of manure is applied, a major mixing can be beneficial. Having a row crop which requires heavy tillage only one or two years out of 4 or 5 years over a rotation will minimize overall soil disturbance.

Some farmers believe weed flushes are needed for good weed control. This practice, working the land on a continual basis several weeks in a row before planting, can damage soils, burn up organic matter, is costly and is not overly effective at getting rid of weeds. A blanket of residues on top, a loose crumbly soil and planting in furrows in fresh clean soil works well on many farms. For weed control, we have tools and understanding today that our forefathers would have died for. Furrow planting, rotary hoes, burners, precision cultivators that go through residues with guidance systems for better accuracy... with these, you don't need to overwork the soil.

Feeding soil life

I have told you that the soil life "wants its food on top and wants to be left alone." But soil organisms, such as earthworms, nematodes, fungi and other micro flora do need food, and raw manure and corn stalks aren't the best diet. Feed the soil life as you would feed your best cow. Sugars and succulent green crops are bacteria foods. Complex crop residues (corn stalks), compost, and humates are fungal food. You need to feed and take care of all soil life.

The plants control who lives and who dies, and so the more plant diversity, the more soil life diversity. Forage blends are helpful not only for the cows but also for the soil. In nurturing soil life you must take every opportunity to grow and provide food for them. I see bare soils as lost opportunities; it is amazing how fast soils can change when given the chance. For instance, to fully nurture your soil life try sowing oats in early fall or early spring, cereal rye grain after corn silage or soybeans, or a year off just for soil building. What works best is a diverse system that introduces checks and balances, allowing no one disease organism to be favored, taking control and causing trouble.

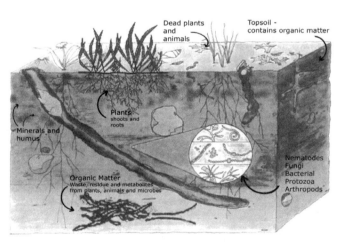

Soil Ecosystem

On our organic farm, as we take on more land that had been conventionally farmed, we use the two growing seasons before it's certified organic to build soils: oats underseeded with clover and grass just cut and left laying buckwheat with clover reseeding itself, cut and left on top, clipping, never letting weeds go to seed and controlling the maturity of the green manure crop.

One method an Iowa farmer followed for building soil life was using only the seeds he had on the farm. He started in the spring with four bushels of oats, working that into the soil in early June. Then he planted three-bushel corn bin run cob corn, working that in when it was four feet tall. Then, in early August, he planted three-bushel soybeans, letting them freeze and shallowly working them in the next year. He got strong crops, very few weeds, and changed his soils in a very short time.

It is easier to change soils than it is to change farmers' thinking and methods. Not everything you do will work out. Timing and weather have big impacts. Organic farmers generally plant their row crops later than conventional farmers, after the soil has warmed up. This makes weed control easier and allows spring grown green manure crops to be utilized.

On our organic farm we aren't afraid to sacrifice a crop. This year our soybeans got weedy, too wet to cultivate, and the stand wasn't great so we turned beans and weeds into a green manure crop. We then planted an emergency forage crop, milo and soybeans, followed by cereal rye. This wasn't an overly profitable year for that land, but we did soil building, didn't allow new fresh weed seeds in, and didn't fight a weedy, poor crop. We don't need a bumper crop on every acre to be successful. In fact, I know of very few farms that get that profitable bumper crop every year on every acre. If 800 out of the 1,000 acres we farm are successful and profitable, and soils on the rest are getting better, our long-term sustainability will be met.

Remember, biological farming is a different system, requiring different thinking, tools and objectives. Getting biological first and then going organic makes the transition easy. Every soil is different (some even have stones), every farm is different; the fun and challenge in farming is to find out what works best in your conditions under your management. You have your own objectives to meet. Visiting successful farms, observing and learning can help you better understand how to meet those objectives.

Soil Testing and Interpreting Lab Results
Glen Borgerding

Soil testing is an important part of a soil fertility program. It is one of the primary monitoring tools used in building and maintaining soil fertility levels. In all farming systems, particularly livestock operations, nutrients need to be monitored and managed for optimum soil, crop and livestock health. In this section we will cover how to take soil samples, factors to be considered in selecting a soil testing lab and how to interpret the results.

Soil sampling

The most critical part in the soil testing process, soil sampling, must be done carefully to accurately represent the fields being sampled. Collecting samples is where most of the variability in soil testing results occurs. There are three popular methods currently used for taking soil samples: grid sampling, zone sampling (usually by soil type) and whole field sampling, which is the most common method used.

Grid sampling requires that a field be divided into grids of 1.5 to 5 acres in size and that each grid is sampled individually. This method is used where zone application of fertilizer is to be done. Advantages to grid sampling are that it allows for localized fertilizer treatments in very small acreages, but it tends to be the most costly in sampling time, lab fees and technology costs.

Zone sampling requires that fields be sampled by soil type within the field. This is generally cheaper than grid sampling but still requires some type of zone application of fertilizer for it to be effective.

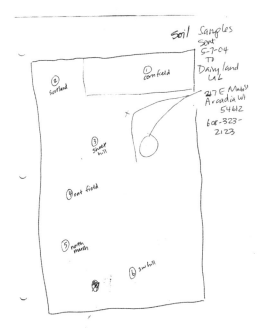

Note where on your property soil samples were taken.

Whole field sampling requires that the entire field be represented by the same sample. The goal is to gather an overall sample that best represents the field or production unit being sampled.

No matter which method you use, it is important to keep in mind the areas in a field that may not be representative of the whole. Old manure or compost pile sites, limestone stockpile areas, livestock feeding sites and field headlands may all pose contamination risks and should be avoided. If an area is responding differently than the rest of a field and is large enough to treat separately, then it should be sampled and tested separately or else it should be skipped over. Other places that should be avoided are areas subject to erosion such as the very tops or bottoms of hills. Use good judgment in soil sampling. Always try to collect a sample that best represents the field.

Soil samples can be taken with a soil probe, or if one is not available, a trowel or spade. Many Extension Agents have soil probes available for loan or rent, or can direct you to a source of a tool. Samples should be collected in a clean plastic bucket or similar suitable container. Do not use copper or galvanized steel equipment or containers as these can contaminate the sample. Sample depth should be between 6 and 7 inches, or what is generally considered the "plow" layer. In grazing or turf situations, a four inch sample depth may be more desirable. Always include the sample depth with the soil sample information. The amount of the sample sent to the lab should be between a pint and a quart unless otherwise stated. Most labs will supply soil sample bags if requested but zip-lock sandwich bags can be used satisfactorily. Make sure to include your name and address and have the field identification clearly marked on the outside of each bag. Additional information such as field cropping history, planned crop and yield goal may be included if you desire the lab to make fertilizer recommendations.

Selecting a lab

Selecting a soil-testing laboratory is a decision that should not be made at random. A farm may have special requirements such as EQIP (Environmental Quality Incentive Program) or CSP (Conservation Security Program) plans through the local NRCS office and/or may be required to maintain a comprehensive nutrient management plan as per local and state agency requirements. In these cases a regional lab may be required, one that utilizes specific procedures or is able to provide information needed for the program with which the farmer is involved. More often than not, the agronomist or fertilizer dealer who works with the farmer may have a preferred lab to recommend and may even take the soil samples and send them in for the farmer.

Lab hopping or using multiple labs over several years should be avoided if possible. Laboratories may vary by region or even by philosophy concerning the types of soil testing procedures they use. This can leave a farmer frustrated with the inability to compare lab results from year to year and from lab to lab. Soil test reporting forms can also vary, causing additional confusion in the interpretation process. This is not to suggest that any one lab is better than another, but that some labs may be more appropriate for a specific region than another lab. If switching labs is necessary, try to select a lab that uses similar methods and extraction procedures so the results are more comparable.

How soil tests should be used

Soil tests have been traditionally used as a tool to sell fertilizer and other soil and crop inputs. A better use for soil testing is as a tool to monitor the nutrient cycling process in the soil. Nutrient levels in soils are not static. Heavy soils tend to change less. Lighter soils that have less clay and organic matter and more silt and sand content tend to change more dramatically. It is this change that needs to be monitored and nutrient levels corrected to truly fine tune a soil and crop fertility program. Many things can change nutrient levels in soils, including crop removal, leaching, nutrients binding up in soil because of high or low pH levels as well as fertilizer additions, manure and compost applications, liming practices or cropping sequences.

How often should soils be tested?

Soil testing in the current era of farming is done rather infrequently. Historically it was considered that sampling every 3 to 5 years was sufficient. This system works well if your soil fertilization philosophy is based on a soil-building program when your soils test low to medium and on a maintenance program based on crop removal if your soils test high to very high. Although this approach appears to be reasonable, it can lead to inappropriate fertilizer and soil amendment applications. A better approach is to sample every 1 to 3 years and to track the changes in the soil tests

Using a soil probe to sample soil.

from sample to sample to monitor that the levels are maintaining or moving in the right direction. This is why using consistent laboratories is so important, so that the results can be compared from one set of samples to the next.

Interpreting soil test results

There are many philosophies on how to interpret soil test results. Some seem over simplistic while others seem too complicated to follow. A thorough explanation of soil test interpretation is beyond the scope of this segment but we will attempt to provide some basics here. It is best to start with a complete soil analysis which includes the primary, secondary and trace elements. When reviewing the soil test report, you will find the results are given in different formats.

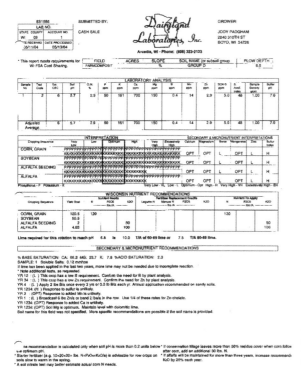

The soil test report offers a lot of useful information.

First are the actual readings of the tests being performed. In most cases nutrient levels are given as "ppm" (parts per million) or in some cases as pounds (lbs) per acre. Multiply ppm by a factor of two to get lbs per acre if conversion is necessary. Some labs also express phosphorus as "lbs per acre" P_2O_5 and potassium as lbs per acre K_2O. If this gets too confusing, just refer to the nutrient level rating system. Most soil test reports have a rating system for nutrients of "low," "medium" and "high." Often, "very low" and "very high" are added. Even though different labs and universities may have different ranges for the nutrient levels that fall within this rating system, the principle is the same. This principle suggests that when a nutrient level falls within the low range, there is a high probability of seeing a positive crop response to the addition of that nutrient to the soil. If it falls within the medium range there is a fair probability of seeing a positive crop response and if it falls within the high range, there is only a slight probability of seeing a crop response to the addition of that nutrient. This allows a farmer at a glance to determine what nutrients should be considered for field application.

Some soil test parameters are not expressed as ppm or lbs per acre. Items such as pH, soluble salts (also expressed as Soil Conductivity or Soil/EC) and organic matter are expressed differently. The pH is a logarithmic value that applies to both pH and buffer pH. Essentially the pH is a measurement of hydrogen ion activity expressed on a scale of 0-14.0, 7.0 being neutral. Below 7.0 is acid, above 7.0 is alkaline. For most crops found on dairy farms, pH readings between 6.5 and 7.0 are considered ideal. Soils below this range should be limed. In soils with a pH of 6.0 or lower, the buffer pH rather than the soil pH is used to determine lime application amounts. The buffer pH test will more accurately predict lime application rates in more acidic soils. It is important to note that the buffer pH is only used for lime requirement determination and should not be considered or used in place of the actual soil pH reading.

The soluble salts test is expressed as millimhos per centimeter or mmhos/cm. This is an electrical conductivity reading that represents the amount of salts in the soil solution. This is one of the few tests on a soil analysis report where it is more desirable to have a low reading than a high reading. Generally a reading of below .5 mmhos/cm is desirable. High salts in soil can cause root stresses and in extreme cases reverse osmosis.

Organic matter is usually expressed as a percentage of the soil volume but some labs may express it as tons per acre. Generally it is considered that higher organic matter levels are better than lower levels. However, extremely high organic matter soils, such as mucks and peats, tend to have unique problems of their own. Coarser soils will tend to have lower levels of soil organic matter. Heavier soils that contain more clay will have higher levels. While all soils should be managed to build and maintain soil organic matter levels, it is important to be realistic. Organic matter levels rise very slowly in soils even under the best management practices.

Another area on the soil test report is the calculated values section. This is where the cation exchange capacity and percent base saturation calculations are reported. While some labs will report these routinely, other labs may only report them if requested. It is important to note that these values are not actual test results themselves but rather are calculated from actual test results. It is a method of summarizing soils in a way that more easily allows interpretation that some farmers and agronomists prefer to use.

In summary, soil tests should be done on a regular basis every 1 to 3 years or at one or more points in the crop rotation. A laboratory should be selected that will fulfill the needs and requirements for the farm being tested. Soil test information should be saved and tracked for each field and compared with yield data as a part of each field's record keeping system. Above all, soil testing should be done to aid in the monitoring of how nutrients are cycling across the farming operation and to verify the effectiveness of the fertility and nutrient management program.

Organic Matter Management, Nutrients and Root Health
Walter Goldstein

Why emphasize organic matter?

Adequate quantities of high quality organic matter are essential for maintaining the beneficial physical functions of the soil (aeration, water holding capacity, drainage), for holding and providing nutrients to crops, and for maintaining soil and root health. Probably about half of the organic matter that was built up under the prairie/buffalo pasture system of the past has been lost by arable cropping on Midwestern farms. The best portions of the original organic matter were mineralized first, and this has caused nitrogen to become less available from the remaining organic matter as it decomposes. In the last half century, maintaining a balance of organic matter in soils did not seem to be important, because nitrogen could be cheaply supplied from mineral fertilizers.

However, maintaining high quality organic matter in the soil is important for all farmers. Even when heavy rates of mineral N (Nitrogen) fertilizer are used, the majority of N that a crop obtains is still probably coming from soil organic matter. Overall soil quality is determined to a large extent by the kind and quantity of organic residues and manures that are used on the farm. Of course, the skillful management of organic matter is especially crucial for the short-term success of farmers, including certified organic farmers, who rely primarily on organic matter sources of N to feed their crops. However, few tools exist to help farmers to understand and better manage their organic matter resources.

The quality of organic matter and the significance of animal manure

The value of organic matter for the soil varies greatly according to the origin, history and quality of that organic matter. Straw, roots, green manures, animal manures and composts all have different effects. Different crops may also have widely different effects on soil organic matter. In general, perennial crops, especially grasses and legumes, can have the greatest positive benefits on soil organic matter and soil quality. This is due to the quality and quantity of their residues. Roots generally provide more stable organic matter than do leaves and stalks of plants.

The quantities of carbon, nitrogen and lignin in organic materials are important for determining their effect on the soil. Proteinaceous materials with high lignin contents are ideal for feeding the soil and crops as well as maintaining or building soil organic matter levels. Probably the most balanced organic fertilizer is ruminant manure because it has both components that release nitrogen to crops and components that are stable and form humus. Numerous long-term trials have shown that, relative to its organic matter content, cattle manure increases or maintains more organic matter than crop residues. Composted manure has a longer lasting effect in building soil organic matter than does fresh manure. Studies with anaerobic manure (such as slurry) suggest that it has a low impact on building organic matter, or even an acceleration of the decomposition of native organic matter.

The transformation of organic matter

Organic matter ages in soil. Young, *labile residues* lose a lot (often 1/2 to 2/3) of their carbon to the air through respiration in a period of weeks or months, partly due to microbial consumption. After 1 to 2 years of decomposition they enter the more stable phase that we call *active organic matter*. These 'middle-aged' organic substances are mineralized slowly by microbes, but they can release a considerable quantity of the N that is obtained by crops. Finally, after about 20 years, the substances enter a new passive phase where they decompose very slowly (turnover of about 1000 years). In this stage they provide very little N to crops, but they do have beneficial physical and chemical effects on the soil. A farmer can, during his or her lifetime, run down or build up the quantity of active organic matter in the soil to the detriment or benefit of future generations.

Organic matter forms soil structure

Labile and active organic matter is physically large (you can see it). Part of it can be sieved out of a soil. It may even look like compost (not a mistake; actually composting seems to manufacture this kind of material!). This kind of 'macro'-organic matter provides a *biogenic (biologically based)* foundation for the formation of large soil crumbs. Such crumbs are conglomerates of organic matter and soil minerals. They may be held together by plant and microbial 'glues' called mucigels. These mucigels may be secreted from the tips of growing roots. These crumbs form and decompose throughout the year. The formation of these biogenic crumbs is also encouraged by mycorrhizal fungi that live symbiotically with roots, and by roots themselves. When these crumbs form, the minerals coat the macroorganic matter and protect it from too rapid decomposition. In fact, if a soil is put back into prairie vegetation it will rapidly accumulate these stable crumbs even more rapidly than it accumulates organic matter. These crumbs have rounded forms, and can be visually distinguished from the sharply angled soil peds formed by tillage.

Earthworms also help create biogenic crumbs. At first, their digestive activity speeds up the decomposition of residues. Afterwards, the decomposition of organic materials in the earthworm casts is slowed down because it is physically protected from microbes by mineral substances. Earthworms multiply where they have access to proteinaceous residues such as manure; they prefer weeds and legumes, and they do not like tillage. They accumulate especially in pasture, in fields that have received animal manure, in fields where leguminous cover crops are utilized or under no-till.

Roots and their importance

When they are young, annual cereal crops put 1/3 to 1/2 of their initial carbohydrate resources into their roots. Root activity includes the growth and turnover of roots, the formation of exudates and root metabolism. Large quantities of roots are formed and die during the growth of these crops, making it difficult to estimate the actual quantities that are formed by sampling at any one time. Furthermore, roots exude substances. These exudates increase microbial activity and directly or indirectly stimulate the mineralization of organic matter and increase the availability of mineral phosphorus in the soil. However, once these cereal plants

Healthy root systems will build organic matter.

begin to form grain they put much less of their resources into roots. Though the formation of water-stable soil crumbs increases while the grain crops are growing, after growth of the crop ceases the stability of these crumbs may decrease strongly.

Having large, active roots is essential for high yields of grain crops. Roots not only take up nutrients, they also send hormonal signals to the shoot (cytokinins) that regulate and encourage photosynthesis and the formation of grain yield. Nitrogen fertilization tends to increase root size. Furthermore, the breeding of wheat and corn for high yields under conventional farming has tended to increase root size. On the other hand, this breeding seems to have reduced the mycorrhizal activity of these crops.

Heavy feeding crops such as corn and cabbages seem to be capable of extracting high quantities of nitrogen from soil organic matter. These crops appear to have a greater ability to extract nitrogen than crops such as small grains or many vegetables.

Results with budgeting real farms

Budgeting results with actual farm systems in three years of on-farm trials suggested that conventional grain cropping systems were depleting their soil organic matter resources. These systems were predicted to eventually equilibrate at organic matter contents somewhere between 2.2 to 3.3 percent, with higher levels where animal manure was applied. The results were about the same for grain crop rotations that used small grains under-seeded with clover as a green manure. However, soils that employed rotations with grain crops and a high percentage of perennial forages were estimated to be accumulating organic matter. Their equilibrium level was estimated to range from 3.9 to 5.4 percent, depending on whether and how much animal manure was applied. The predictions made by the budgeter on how different farming systems would affect the equilibrium soil organic matter contents were reflected in the actual soils on the different farms. The cash grain systems generally had lower organic matter contents.

See Appendix A for more information on organic matter research results.

Farmer Profile: Wayne Peters - Chaseburg, WI

Jody Padgham

Certified organic since 1988, Wayne Peters is one of the "old guys" involved in organic dairying. Now milking about 110 cows and using 450 acres, he was one of the founding fathers of the now over $200 million cooperative CROPP/Organic Valley. Wayne's sons now farm with him in the hilly land of southwest Wisconsin. The Peters feel good about their 22,000 pound herd average and 8 percent butterfat from their Holstein herd.

"I like to say that I got into organics out of 'cheapness,'" Wayne says. "I grew up before they used chemicals to farm. I hated how much it cost to spread those fertilizers and herbicides, so I decided to quit." Wayne notes that his sons have had the opposite experience; they have been raised with an organic system. "They complain about what it costs to run the flame weeder for a day, but they have no comprehension of how cheap that is in comparison to buying chemical controls." Wayne notes that organic farming means fewer inputs and less input costs, but more labor time.

Over the years Wayne has developed a strong expertise in crop management and soil fertility. He was working for a silo company when he first started thinking about organic production, and learned about the benefits of injecting manure below the ground surface. At the Peters' farm they use a gutter cleaner to bring manure out to a pit all winter long (cows are inside a good part of the day in the winter).

The manure develops a crust and ages in the pit. In the spring the slurry is pumped into 3000 gallon tanks and injected 4 to 10 inches under ground. Wayne feels that this system of returning nutrients to the ground has been instrumental in the increase in soil fertility he has seen over the years. "I can see the ridge soils actually getting darker," he says. "That is the added organic matter and nutrients from the years of manure injections." Wayne also uses Midwestern Bio-Ag consultants, and adds trace minerals and potash as needed.

Wayne is a strong advocate of strip cropping, and their farm was one of the originals to work with the practice when it was first developed back in the 1930's. On his very steep hilly ground, Wayne's system is to plant his strips in strict accordance to the contours of the land. "By starting with a 60 foot strip, and staying with the contours, you develop a terrace-like effect." This allows runoff water (and nutrients) to be caught at regular intervals, again preserving the integrity and fertility of the soil.

For weed control, Wayne relies on cultivation. "A major difference I see between organic and non-organic systems is the amount of cultivation." Cultivation on the Peters' farm is well timed and frequent. "It is harder in a wet year like this has been," Wayne says. "We needed to get in early when it was still wet; we had to go anyway, otherwise we would have really had troubles." The Peters have a flame weeder, and use it every year to set back the early weeds. "It is always a guess as to when to go out with it. This year I think we could have gone out into the corn earlier."

"Even though I've been doing this a long time, I still have to learn new things all the time," Wayne notes. For example, in the past few years he's decided to stop growing as much corn and instead bring on more small grains. "I used to rotate with oats, but now have tried winter wheat and barley. The wheat will overwinter. I feed it to the chickens and steers. The barley goes to the cows." His rotation is two years corn, one year small grain, two to three years hay. "The small grains mean new types of weeds, new management, new things to learn about how they do on my soils."

"I believe that no one should experience an overall drop in production with an organic system," Wayne says. "You will be forced to cull out the sickly cows and poor quality cows that you were able to limp along using drugs. Judicious culling will leave you with an overall strong and healthy herd, good genetic stock, fewer hereditary diseases (like mastitis tendencies). I see organic as a way of fine-tuning your herd, looking for the best to fit your particular system."

Wayne offers tips for keeping down flies in the barn: 1.) Keep fresh manure out of the barn. 2.) Put natural fly strings up early and change them as needed. 3.) Try several 3-4 foot fans constantly going in the barn — they create a 3 mile per hour breeze and "suck the flies right off the cows."

Wayne's suggestions to those starting out? "I have found that by improving the soil, I can raise healthy crops and then produce healthy cows and a quality product. I recommend farmers really learn to understand their soils."

Chapter 10

Organic Cropping Systems

Cows need to eat, and unless you plan to buy all your feed, a start up or conversion to organic dairy production is as much about understanding organic crop production as it is about a new kind of animal husbandry. The organic farmer must rely on crop rotations, cultivation and other non-chemical management tools to maintain healthy soils and high quality feeds for a dairy operation. In this chapter we discuss tools and techniques critical for successful organic cropping systems.

Understanding Crop Rotations
Mary-Howell Martens

Many American crop farmers grow two crops--corn and soybeans. To organic farmers, this is not crop rotation. Indeed, this is what one certifying agent calls "crop oscillation." Despite mesmerizing images of crops endlessly oscillating up and down, tall and short, organic farmers can't get away with this approach to crop rotation, because it does not create a healthy, sustainable environment on our farms.

Organic production requires a diverse crop rotation with a wide variety of crop types. The NOP standards state "crop rotation must maintain or improve soil organic matter content; provide for pest management in annual and perennial crops; manage deficient or excess plant nutrients and provide erosion control." (NOP 205.205). Crop rotation is indeed our best defense against insects, diseases, weeds; and it is the optimal way to manage soil fertility and increase organic matter. Fortunately, organic markets usually reward crop diversity, making it possible to plan healthy, long-term crop rotations that improve soil and avoid serious pest problems.

To plan crop rotations, we must consider long term and whole farm effects that address the following questions:

1. Which crops are agronomically well-adapted to my soils and climate and will maintain and improve the long term productivity and health of my soil?
2. Will my intended crop rotation control erosion, minimize pest damage and disease, break weed cycles and add organic matter to the soil?
3. Will my intended whole-farm crop rotation produce a consistent and adequate income over multiple years by producing a variety of crops that supply my animal feed needs or have a reliable market and price?
4. Will my intended crop rotation make effective use of my available resources, including labor, time and equipment?

Each type of crop has a distinct effect on the soil, although exactly what the effect is and how it influences the growth of subsequent crops has not been well studied. Some crops have high fertility demands, others produce much organic matter, while others leave excess fertility behind. Many plants release substantial amounts of sugary exudates from their roots, which stimulate the growth of a signature population of microbes. A special type of beneficial root fungi, called mycorrhizae, actually grow into plant roots and feed off plant sap while providing water and nutrients to the plant. It is undoubtedly due in part to these characteristic "farmed" microbe populations that some crops do better following certain crops than others.

Crop rotations: the basic principles

Crop rotations must address both the immediate needs of each crop and the long term productivity and health of the soil, the farm and the farmers. Every region has a group of crops best adapted to grow there, so it is difficult to make actual rotation suggestions. However, in general, good rotations should:

1. Alternate legumes with non-legumes to provide nitrogen, but not create situations where excess unused nutrients can leach.
2. Alternate row crops with solid seeded crops, and fall-planted and spring-planted crops to break pest and weed cycles. Organic farmers can be creative with crop choices — some dairy farmers have been successful growing highly nutritious root crops and brassicas, such as turnips, fodder beets and kale, for high quality animal feed late in the season.
3. Use cover crops and other crop residue to add active organic matter. Soil organic matter is critical because it feeds microbes and other soil life, cycles nutrients, holds water, improves soil tilth, and decreases erosion and leaching.
4. Employ deep shading crops and allelopathic (plants that inhibit the growth of other plants) crops where extra weed control is needed.

5. Incorporate crop mixtures for more diversity. Because organic farmers don't contend with herbicide residue carryover, crop mixtures like barley/pea/oats, triticale/peas or soybeans/sorghum are possible. Especially for dairy farmers, these mixtures can produce profitable yields of valuable forage or mixed grains, and are well adapted to some areas or season niches where more traditional row crops, like corn and soybeans, might not be successful.
6. Alternate deep rooting crops with shallow rooting crops to help keep soil structure open and assist with drainage.
7. Avoid bare soil during the winter and incorporate reduced tillage when practical for erosion control.
8. Spread out peak labor demand over the year by staggering planting, cultivating and harvest of a variety of different crops and crop maturities.
9. Plan for relatively consistent acreages of different crops each year for stable long-term income, on-farm use and marketing.

Oats underseeded with clover provide multiple benefits.

Organic farmers need to experiment with different rotations and learn which will work well on their farm. For a beginning organic farmer, it would probably be best to chart the crops planned for all fields over the next 3-4 years. This will help to visualize long term rotations on individual fields while considering the overall balance of crops on the entire farm.

It is important to build a rotation with crops that work with available equipment, labor, time, storage and markets. Try to avoid planting more of any one crop than can be handled in a timely manner. Row crops will require considerable time in the spring for tillage, planting and cultivating. Getting the timing of these operations right is extremely critical. If you don't have labor enough to insure that you will be able to complete these operations correctly, it would probably be a better idea to plant more of your acres to small grains or hay which peak in labor requirements at a different time of year.

The economics of crop rotation must not be ignored. While planting the whole farm to soybeans one year and the whole farm to corn the next is indeed an organically acceptable rotation, it is certainly not the best approach to deliver a consistent supply to buyers or your animals, or to ensure a consistent annual income for your farm. Many buyers prefer to develop relationships with farmers on whom they can rely for a certain quantity and quality each year. When you wish to expand your repertoire of crops for better rotations, you may need to locate or develop new markets.

Sustainable crop rotations

It is important to recognize that a crop rotation that works well for five years won't necessarily work as well for twenty years. Dr. William Albrecht, professor of soil science at the University of Missouri College of Agriculture in the 1950's and 1960's, studied experimental plots of different crop rotations over many years. He found that when high yield was the primary goal, even when a reasonable rotation was followed but without adequate additions of mineral fertilizers, soil fertility declined and yields dropped.

In contrast, we know that organic vegetable farmers, growing crops intensively on very small acreages and supplementing heavily with compost and manure, can quickly build up excessive fertility, especially nitrogen and phosphorus, leading to increased weed, insect, disease and leaching problems. This can also happen on dairy farms if there is insufficient land for manure application. While both approaches technically meet organic standards, neither is sustainable or productive over a long period of time.

Crop rotation must always be an evolving process on an organic farm, matching crops to ever-changing soil conditions, animal needs, market demands, weather fluctuations, and labor and machinery limitations. We should re-evaluate our rotational strategies frequently, taking soil tests every 3-4 years to track changing nutrient levels, and reconsider whether our current crop rotation approaches are meeting our desired goals.

Precision Organic Farming
Mary-Howells Martens

Organic farming is not excessively difficult, but it does require the farmer to be a better farmer, a more observant farmer, a farmer who practices more careful agronomy and more conscientious animal husbandry. That's the way the organic system works. We don't have chemicals to patch up mistakes, so we have to be very careful to "do it right" the first time.

In precision conventional agriculture, people think of big equipment, GPS positioning, genetic engineering and powerful chemicals. We contend that organic farming, regardless of the size of the farm or the equipment, is actually far more precise than most conventional precision farming. Electronics can be useful, but they will never take the place of the farmer's brain, eyes and experience.

Precision observation

There is an old saying, perhaps largely forgotten in this modern era of on-board computers and climate-controlled tractor cabs, that the best fertilizer for a field is the farmer's footprints. Nothing quite equals your own observations and experience. There is no electronic replacement for knowing your fields, really looking as you work them, making notes of where there are particular weed problems, crop growth differences, fertility imbalances, drainage problems or big rocks. The real heart of precision farming is observation--YOUR observation--and your ability to figure out what those observations mean.

It is important to make conscious connections between what you do in a field and what the results are. As organic farmers, we remember where a field was plowed too wet because the soil damage can last for years, or where the cover crop was particularly lush because this may have added extra nutrients and microbial diversity. When yields are particularly high or disturbingly low or when certain types of weeds proliferate, we must try to make connections between what happened 6 months, 1 year, even 10 years ago, to what the outcome now is.

Precision equipment operation

We could spend a lot of time talking about precision adjustments for plows, tillage implements, grain drills and combines. Careful and conscientious attention to repair, adjustment and use on every piece of equipment will pay off in higher yields and quality.

The adjustment of the corn planter is a good example of the importance of precise equipment operation, though the same amount of detailed attention should be given every other machine and operation on the farm.

The corn planter is a complex machine with many moving parts that must work together precisely for optimal seed placement and coverage. Before the season begins and as planting proceeds, thoroughly check whether the corn planter is level and if all parts are properly aligned, not worn and in correct adjustment. Worn chains and sprockets may be working improperly long before they actually break. Coulters, firming wheels, and depth wheels must be set correctly for accurate planting. Metering units should be taken apart and cleaned. The planter should be checked for broken fingers and brushes. The seed belt should be checked for cracks and flexibility. During planting, check seed placement and seed opener disks to determine whether the seed is falling and staying where intended.

Retrofitting with shoes, firming points, specially designed seed tubes or "eccentrically" (off center) bored gauge wheel bushings will often result in more uniform seed placement than what the planter had when it was new. Trash wheels in front of the gauge wheels can sweep away clods and stones, making for a level surface and therefore uniform planting.

Art Scheele of the American Organic Seed Co. says that of all the money spent producing a crop, quality planting is THE best investment to give a good return. Differences in planting corn can result in a 3-8 bushel/acre spread in yield, and in a poor year, the difference can be even greater. Too often, precision planting is one of the last things farmers think about. Frequently, planters are not maintained, worn parts are not checked and replaced, farmers try to run their planters too fast, they don't calibrate them under real field conditions, and rarely do they dig in the soil to see whether the seeds are actually being planted correctly.

Seed placement is probably the most critical factor in planting, whether you are using a fancy state-of-the-art corn planter or planting by hand in the garden. Seeds of all species have an optimal depth at which they germinate best, depending on soil conditions. Differences in depth can result in a 20 percent difference in yield and have a profound effect on time of emergence and seedling vigor and health.

A seed planted too deep will emerge slowly and is more vulnerable to fungal rots and insect attack. A seed planted too shallow will not form an adequate root system. A half-inch difference in depth can result in seedlings emerging one day later, more if conditions are poor. Seeds placed into cloddy soil or not into moisture won't grow until it rains. A corn plant emerging significantly later than its neighbors will never grow as well. Its growth will be suppressed by larger plants, and it can essentially become a weed in the field.

Non-uniform emergence makes for difficult weed control. Since timing is everything when it comes to effective mechanical weed control, crops that emerge vigorously and uniformly will make our timing of all subsequent field operations more effective. Organic farmers must do all they can to achieve uniformity and vigor in seedling emergence.

Precision timing

Sometimes you can learn more from a bad situation than from when things go well. The year 2000 was like that for us. As the spring moved into summer, the rain just wouldn't quit for days on end. In a frenzy to get the corn planted and weeded in the few clear days between downpours, my husband worked later and later at night, sometimes to three o'clock in the morning or later, planting into soil he knew was not fit.

This siege pressure did result in most of the acres getting planted as planned, but the stress on everyone was enormous and harvest time showed us something even more discouraging. Sheer determination to get the seeds in the ground had not been good enough. Yields were low, and long-term damage to our carefully tended soils was painfully evident.

All crops have an optimal time when they should be planted depending on region and variety. Once you get out of this optimal planting window, plant vigor and yields will suffer. For example, in New York, barley should get in by the first week of May, oats should be planted by May 15, corn should be all in the ground by June 10.

If the ground isn't fit or if the particular crop is outside its optimal planting window, we may be better off moving on to something else, even if all the intended acres are not yet planted. If we focus exclusively on getting corn planted, we might not get the soybeans or the other crops that follow planted during their most favorable planting windows and the weeding and cultivating may not get done on time. We are far better off taking a more holistic view of the farm, striving to keep the timing of planting, weeding, cultivation and harvest of all crops as close to optimal as possible.

The core of making precision timing effective is the idea that diversity spreads out risk and labor. If you are only growing corn and soybeans, then corn and soybeans must go in, regardless of the weather. Organic farmers generally have considerably more crop diversity, and there are good markets for a wider range of organic crops. The more crops you have on your farm, the more likely you are to hit the proper planting window correctly for each crop, get the weeding done, and spread the work out to avoid stressful labor crunches. A diversity of crops also spreads risk — the more you have, the more likely it is that at least some of the crops will thrive, regardless of what the weather brings.

Precision farm functioning

Sometimes it rains. Rainy days are just as necessary in the precision organic farming process as the warm sunny ones. This is the time to catch up on necessary record keeping, rethink planting plans, repair and maintain machinery, talk to your banker and keep her up-to-date about what is happening on the farm. Talk to your seed supplier and crop insurance agent. Get re-organized and re-prioritized. It is a good time to check labels and be sure that all your inputs meet organic standards.

Rainy days are also a good time to relax, spend time with family and friends, and to recognize that stress management is absolutely necessary, but is all too often neglected on a farm. Farming is a great way of life . . . but only if it truly is a great way of life for everyone involved. For precision organic farming to be successful, all members of the farm must be working together, appreciating what everyone else is doing to contribute to the success of the farm and wanting to be there. If this isn't the case, then all the GPS and computer monitors in the world won't help.

But when it is, precision organic farming hums just like a well-oiled machine (in good adjustment and repair!)

Pest Management in Organic Cropping Systems
Elizabeth Dyck

Crops pests consist of weeds, insects and the pathogens (chiefly, certain types of fungi, bacteria, viruses and nematodes) that cause plant disease. Worldwide, these pests are estimated to reduce crop yields by at least one-third annually. In U.S. conventional agriculture, misguided attempts at control through the use of pesticides cost billions of dollars each year, have actually increased pest problems in many crops and have caused long-term damage to natural resources. Organic farming systems are based on principles that reduce the risk of major pest outbreaks. However, even the best-run organic farms are not immune to crop pest attack. Adverse weather conditions, influx of pests from neighboring farms and naturally occurring populations of crop pests can lead to substantial crop yield loss on organic farms.

This section will first review the underlying principles that provide the best protection against major outbreaks of all three types of crop pests. A brief consideration of management practices specific to weed, insect and disease control follows.

Principles of organic pest management

Successful pest management in organic cropping systems rests on three major strategies:

Good soil management

Sound soil management is in fact the most important strategy underlying successful pest management in organic cropping systems. Soil that is maintained in good tilth and with adequate levels of fertility leads to healthy, vigorous crops that are more resistant to attack from pests. In contrast, crop plants grown in compacted or eroded soil (resulting in constricted root systems) or grown with unbalanced fertility (either deficient or excessive amounts of essential nutrients) can be more susceptible to attack by insect pests and diseases. Weak crop plants are also less competitive with weeds. Moreover, soil management that avoids excessive tillage and provides steady and diverse inputs of organic matter (through crop rotation, green manuring, cover cropping, application of compost and/or judicious use of animal manures) promotes diverse and abundant populations of soil organisms. Such populations contain organisms that can actually reduce the number of disease pathogens by 1) competing with them for essential nutrients and other resources, 2) inhibiting their growth and development through the release of toxic compounds and 3) just plain eating them.

Creation and maintenance of diversity on the farm

Diversity is fundamental to crop pest management in at least two ways. First, maintaining a diverse mix of crops on the farm "spreads the risk" against major crop loss to pest attack. Although a few pests can attack many crops, none can devastate a combination of annual, perennial, grass and broadleaf crops. Second, maintaining diverse environments on the farm can also reduce the risk of pest outbreaks by providing habitat for beneficial organisms that kill or parasitize crop pests. Cropping practices that promote a diverse "farmscape" include rotation, intercropping, strip cropping and cover cropping. Diversity can also be enhanced through keeping a portion of the farm in permanent pasture and by plantings of non-crop species (including shrubs or trees) either alongside fields or in strips within them.

Understanding the biology of major pest species

By becoming familiar with pests' biology — their life cycles, what weather, soil, and management conditions favor them, what crops and non-crops they feed on, and what organisms feed on them — farmers can minimize pest problems in two ways.

First, they can avoid the development of conditions that are favorable to pests. For example, consider the prohibition on use of treated seeds by the National Organic Program. This prohibition initially worried some organic farmers who considered treated corn seed a "necessary evil" for protecting against seed and seedling rots. However, most of these diseases are favored by cool, wet soil conditions. By delaying corn planting until the soil is at or above 55°F and sufficiently dried out to permit field operations without compaction, farmers minimize the risk of these diseases.

Pests on soybeans may be controlled with crop rotations.

Second, an understanding of pest biology helps farmers manage pests that are already present by identifying "pressure points"— aspects of the pests' biology that make them vulnerable to attack. Take the perennial weed quackgrass, for example. Quackgrass survives and thrives in farmers' fields from year to year largely because of its rhizomes — the "dead white fingers" of stem that run underground and from which new shoots spring. Quackgrass rhizomes can actually benefit from spring cultivation since it tends to break them up into smaller pieces — each of which can produce new top growth — and distributes them under cool, moist conditions that are optimum for their survival and growth. However, quackgrass rhizomes are very susceptible to desiccation — therefore a tillage or cultivation operation done in midsummer that exposes rhizomes to hot, dry conditions at the soil surface can help fight quackgrass infestations. (For sources of information on pest biology, see the references section.)

Some Really, Really Important Considerations!

Underlying the three general principles listed above are assumptions that differ radically from those underlying the conventional approach to pest management. By keeping the following points in mind, organic farmers can avoid falling into the "substitution trap"--substituting "organic" inputs for conventional pesticides--that inevitably result in higher costs and growing pest problems.

Pest management in an organic system

- emphasizes the prevention of pest buildup rather than control methods once pest outbreaks have occurred. "Natural" sprays and other pest management inputs that are acceptable to certifiers may be useful in certain emergency circumstances, but organic farmers spend the bulk of their efforts and resources on the creation of a healthy environment for crop growth.

- accepts that pests cannot, and should not, be eradicated. Pest species have vital roles in the farming system. Weeds provide habitat and food for beneficial species and protect against soil erosion. Insect pests, bacteria, fungi and nematodes themselves serve as food for beneficial species and aid in the breakdown of plant tissue--an essential step for nutrient recycling.

- works through an integration of many strategies, rather than relying on a "magic bullet." In the case of quackgrass, for example, the use of midsummer tillage as described above by itself is unlikely to substantially reduce quackgrass infestation. However, when it is used in combination with other strategies, for example, smother cropping, interseeding and crop rotation, quackgrass can be knocked back on its heels.

- takes work and study and must be tailored to individual farming systems. For example, gaining and using knowledge of pest biology will require some research as well as observation of the pest in the field. Before adopting a practice, organic growers must also consider its possible negative side effects as well as whether it will work under their farm's soil and climatic conditions. Small-scale experimentation with a variety of methods is recommended before adopting a practice for your entire system.

Weed management

Managing Crops To Suppress Weeds
In addition to the three general principles of pest management outlined above, organic farmers have a fourth major "weapon" to use against weeds - their crops. The following cropping practices actively suppress weeds.

- A diversified crop rotation is a fundamental tool for weed management. Crop rotation should include three diverse crop types: row crops, smother or cover crops, and sod crops. As shown in Table 1, inclusion of these types of crops in a rotation subjects weeds to multiple stresses and to an unstable environment that prevents their buildup. A 13 year study in Lamberton, MN has shown that an organic rotation containing these crop types (in this case, a four-year rotation of corn-soybean-oat/alfalfa-alfalfa) by itself reduced annual weed density by over 50 percent in comparison to an organic rotation with just row crops[20].
- Crop planting dates should be optimized to favor the crop, not the weeds; cool-season crops like small grains should be planted as early as possible. A late-planted small grain has lost the competitive edge against summer annual weeds like foxtail and pigweed. Warm season crops should not be planted until the soil has warmed to over 50° F so that successive flushes of spring annual weeds like wild mustard and lambsquarters can be destroyed through preplant tillage.
- Short-term cover crops of buckwheat, sorghum-Sudangrass, or Japanese millet can knock back weeds (including perennials like Canada thistle and quackgrass) in problem fields or in weedy patches in fields. Care needs to be taken to mow or turn under these crops before they flower to avoid the problem of volunteer plants.
- Cover crop mixtures that form a dense living mat over the soil, for example, intercrops of winter rye and hairy vetch, forage soybeans and Sorghum-Sudan grass, or field peas, oat and alfalfa, are especially effective at smothering weeds.
- Fall cover crops like winter rye are not only valuable for soil management, but have the potential to suppress weeds early in the next season through an allelopathic effect - release of chemical compounds that inhibit weed germination and growth especially of small-seeded weeds.

Buckwheat helps to suppress weeds between crops.

---

[20] *Elizabeth Dyck, 2003. Unpublished research results, University of Minnesota's Southwest Research and Outreach Center.*

- Narrow row spacing in small grains and other drilled crops accelerates crop canopy closure and reduces the spatial "niche" available for weeds.
- Seeding crops at the maximum rate optimizes crop competitiveness with weeds. Crops that demand high fertility like corn constitute an exception to this rule. For corn, a more moderate planting rate will generally better match soil fertility conditions.

Mechanical Weed Control In Row Crops

Mechanical cultivation is another important tool for weed management in organic systems. Unfortunately, in some organic systems, mechanical control has been over-emphasized and used as a substitute for herbicide use. Over reliance on mechanical control not only has negative effects on the soil, but ultimately increases problems with weeds that are adapted to its use.

Many organic producers like the rotary hoe.

In a collaboration of organic and converting farmers with researchers, three years of experimentation on mechanical cultivation techniques in corn were conducted in certified organic systems in Minnesota[21]. The major conclusions of these studies suggest some practical guidelines for use of mechanical weed control in organic row crops:

- Early season cultivation (during the period from time of planting to inter-row cultivation) is important to reduce weed infestations and avoid crop yield loss to weeds. Inter-row cultivation used without early season control methods resulted in significantly higher weed densities and increased the risk of yield loss to weeds.
- Type of equipment used is not a critical factor in successful early season cultivation. Rotary-hoeing, flex-tine weeding, flame weeding and even a spike-tooth drag worked equally well in terms of weed suppression when followed by inter-row cultivation. Use of rotary hoeing, flex-tine harrowing, and flame weeding also all maintained crop yield at the level of a hand-weeded control.
- A limited number of targeted or well-timed cultivations are just as effective as more frequent cultivation in controlling weeds. Increasing frequency of rotary hoeing from two to up to six times per season did not result in either decreased weed density or increased yield. (In the targeted approach to cultivation, daily scouting of the field is used to monitor weed germination--

---

[21] *Ibid.*

appearance of white roots underneath the soil surface. When weed roots are visible in moderate to high numbers and soil conditions permit, the field is cultivated.)
- In addition to the timing of cultivations, another critical factor for successful mechanical weed control is adjustment of both early season equipment and the inter-row cultivator--to provide maximum "weed killing power" without negatively affecting crop plants.

Preventing Weed Seed Production and Dispersal
Hand-weeding has a bad reputation among field crop growers, but can have a critical role in preventing serious weed infestations. After last cultivation, fields that still have isolated weed plants or small patches of weeds can be walked to pull the weeds, thereby preventing further inputs of weed seed into the soil. In cases where--whether due to adverse weather, grower miscalculation, or just plain bad luck--weed infestation in a field is high and no more control methods are available, growers need to consider the long-term consequences of letting the crop go until harvest and thereby allowing the weeds to set and release seed. It may be possible to harvest the crop for forage rather than grain--or it may be wisest to disc the crop under, gain the benefit of the green manure and try another crop later in the season or next year.

Be sure purchased seed has low weed seed content (upper label).

Insect Pest and Disease Management
The following practices aid in disease or insect pest management--or in some cases both. These practices are most successful when farmers combine them with an understanding of the biology of the major disease pathogens/insect pests of their farming region.

Crop Rotation
Among the many benefits of crop rotation is the role it plays in protecting against crop disease. Rotation reduces the risk of diseases in which the pathogen survives in the soil or on crop residue. An important rule of thumb for designing crop rotations for disease control is alternation of grass and broadleaf crops. Disease organisms tend to attack grasses or broadleaf crops but rarely both. For example, Fusarium fungi attack both small grains (causing scab) and corn (causing root, stalk, and ear rots). Rotation to broadleaf crops can reduce the amount of Fusarium in the soil. Similarly, levels of the fungus Sclerotinia, which causes white mold in many broadleaf crops, can be reduced by rotating to small grains, corn, or other grasses. Certain crops, for example oats, buckwheat, and millet, appear to be less susceptible to many of the major diseases affecting both grass and broadleaf crops and therefore can serve as "general purpose" break crops in the rotation. Crop rotation can also protect against some insect pests, for example, northern and western corn rootworm.

Clean Seed

Certain disease pathogens, e.g., soybean mosaic virus, can be spread through infected seed. Improperly cleaned seed can also spread disease, as in the case of soybean seed that contains soil contaminated with soybean cyst nematode. Farmers should use well-cleaned, high quality seed from fields or sources known to be disease free.

Use of Resistant Varieties

When possible, growers should make use of resistant varieties that have been developed with traditional breeding techniques (that is, without the use of genetic modification techniques, which are not allowed in organic systems). Increased cooperation between breeders and the organic community would increase the number of resistant varieties whose seed is available as certified organic.

Sanitation

Washing field equipment before moving between fields can avoid the spread of pathogens throughout the farm. Some disease pathogens (e.g., those causing potato late blight, wheat scab, brown stem rot of soybean) and insect pests (e.g., European corn borer and southern corn rootworm) can survive on crop residue. Burying crop residue in the soil can reduce the survival of such pests. Clean cultivation to remove all weedy growth two or more weeks before crop planting can also be effective in managing certain pests, e.g., cutworms in corn. However, the use of clean cultivation for pest control should be balanced against its negative impact on soil quality and used only sparingly and in conjunction with soil-enhancing strategies. For example, plowdown of main crop residues should be followed with planting of a cover crop.

Crop Mixtures

Intercropping can be an effective tool in managing insect pests. A common farmer practice in planting legume green manures, for example, is to plant a mix of species, say, alfalfa, sweetclover, and red clover. This particular mix ensures a legume crop even if potato leafhopper or sweetclover weevil attacks. Intercropping may also help protect a crop against insect attack, e.g., intercropping alfalfa with a grass can reduce potato leafhopper and alfalfa weevil damage.

Strip Plantings

Organic vegetable producers have long grown patches or strips of plants whose flowers produce abundant pollen and nectar to attract insect species that are natural enemies of insect pests. Plants such as dill, coriander, buckwheat, tansy leaf, and many others have been shown to attract beneficial insects. Research suggests that field-scale planting of such species (as strips within a field or along its edges) can help manage crop insect pests. However, careful design is needed. For example, plant species need to be chosen that flower at the appropriate time, attract the specific beneficial insects needed to control major pests, do not attract other crop pest species and are not a danger of becoming weeds. Strip plantings of crops can also help control insect pests. Strips of alfalfa, for example, have been shown to provide habitat to hundreds of insect species, including many beneficials.

Manipulation of Planting/Harvesting

Field operations can sometimes be timed to avoid peak populations of crop pests or increase beneficial species. For example, delayed crop planting can reduce bean leaf beetle damage in soybeans, and, in the northern tier of the Midwest, European corn borer attack in corn. Strip harvesting of alfalfa, which provides beneficial species with continuous shelter and food, has been shown to dramatically increase the number of natural enemies of the major insect pests of alfalfa.

Soil strips

Purchased Inputs

A wide variety of commercial pest management products are now available to organic growers, including natural enemies reared and shipped for mass release, "botanical" pesticides derived from plants (for example, pyrethrum and various formulations from the neem tree), and "biopesticides," including products containing the bacterium *Bacillus thuringiensis* (Bt) or the fungus *Beauveria bassiana*. However, application of these products is usually prohibitively expensive in field crops, and their use can have unintended consequences, including destruction of beneficial species and the development of resistance to the input by the pest species.

In dropping dependence on pesticides and herbicides organic farmers in no way lose control over their production system. There is a wide diversity of tools useful in managing for healthy, weed and pest-controlled crops. Through planning and careful management using the methods we've listed, a high value organic crop can be produced.

Table 1: Essential crop types for a weed-suppressive rotation

Row crops	Smother/cover crops	Sod crops
Examples: corn, soybean, sunflower, dry beans	Examples: small grains, buckwheat, Sorghum-Sudangrass, Japanese millet, fall brassicas, grass/legume mixes (e.g., oat/pea/vetch; small grain/hairy vetch)	Examples: alfalfa, red clover, sweetclover, forage legume mixes, legume/perennial grass mixes
1) Wide row spacing permits physical removal of weeds through cultivation or hand weeding through midseason 2) Later planting date than small grains allows preplant tillage to stimulate flushes of weeds that can then be killed through cultivation	1) Compete with weeds for light, nutrients, and other resources due to high density planting broadcast planting or narrow-row spacing 2) May exhibit allelopathic activity against weeds (e.g., winter rye, wheat, oats, barley, buckwheat, Sorghum-Sudan) 3) Can be interseeded with a legume (with little or no yield loss) to intensify weed suppression and add to soil improvement 4) Often not full season crops—allow variation in planting date and tillage, e.g., a) early spring planting/ late summer tillage (small grains) b) tillage throughout spring/early summer planting (buckwheat) c) late summer tillage/fall planting (winter rye or wheat)	1) High density planting with nurse crop out competes weeds 2) Lack of soil disturbance for two or more years kills a significant fraction of buried weed seed (especially grass seeds) 3) Mowing and grazing are tolerated by sod crop but keep weeds in check by preventing seedset 4) May exhibit allelopathic activity, e.g., alfalfa, red clover 5) Soil inversion (plowing) at end of the sod crop weakens perennial weeds, e.g., Canada thistle, quackgrass 6) Creates habitat for organisms that attack weeds, e.g., carabid beetles that eat weed seeds 7) Legume sods function as slow-release nitrogen sources that may favor crop over weed growth

Chapter 11

Pasture Management

The NOP requires that ruminants have access to pasture, with pasture defined as "§205.2 Land used for livestock grazing that is managed to provide feed value and maintain or improve soil, water and vegetative resources." For many farmers this is fulfilled by allowing cows out on a grassy paddock every day. The NOP does not state specific stocking numbers, due to variations in pasture quality and other factors. A certification agency in your area can help you understand basic pasture requirements for organic certification and cow health.

Many organic farmers are finding that moving to a management intensive grazing (MIG) system not only increases the quality and quantity of forage per acre, but lowers inputs and increases farm profitability. We discuss here the basics of pasture management, with a MIG focus. The basic management techniques used here are also applicable to non-MIG pastures.

Pasture Management
Laura Paine

To successfully convert to a pasture-based dairy system, it is valuable to make a simple but fairly significant shift in thinking: stop viewing pasture as a crop to feed cattle and think about using cattle to manage the grass. Farming is all about capturing the sun's energy and converting it to usable products. If we start by maximizing the amount of energy we capture (focusing on the grass), all other steps in the process have greater potential to yield profits.

The first question many people ask about pasture management is about what species or varieties to plant. Many grass-based dairy farmers spend a lot of time and money replanting and renovating pastures that don't seem to be performing well. In reality, the species planted is far less important than how the pasture is managed. There is no perfect plant or combination of plants that will make a pasture system work. The best results are often gained by working with what is already there.

In fact, the most important concept in management intensive grazing (MIG) is management. Success requires an intimate knowledge of one's land and its capability, gained through daily monitoring. The inputs needed to make it work well are observation skills, creativity and adaptability.

Pasture ecology

At the heart of organic farming is the premise that agriculture should be modeled on the natural ecosystem and in doing so, we hope to harness nature's inherent checks and balances to control pests and enhance crop growth. The more closely we can mirror natural ecosystem functions, the better our farming system will work.

Of all the crops that an organic farmer can grow, grazing ruminants on pasture comes closest to that ideal marriage of production requirements and ecological function. Natural ecosystems inevitably move over time from simplicity to complexity and stability in a process known as succession. Annual cropping systems reset the succession clock and revert a field back to the beginning each year. The weeds and pests we battle are the early successional species that colonize the open habitat we create with the plow.

Simple electric fence systems create managed paddocks for grazing.

The perennial nature of pasture systems allows succession to proceed farther along the road toward the stability characteristic of natural grasslands. Even the simple plant community of a pasture is inherently more stable than an annual crop rotation. The richer and more complex the pasture plant community is, the greater its ability to support the diversity of animal, insect and microbe species that make organic systems function effectively.

Grazing animals evolved as an integral component of grasslands. The key to effective pasture management is harvesting the products of these animals with minimum disturbance of ecosystem processes. With good management, a pasture just gets better and better over time.

Maximizing pasture productivity

Fortunately, the same practices that enhance ecosystem processes also lead to improved pasture productivity. Management intensive grazing involves understanding four basic elements of grass management: rest, intake, residual, and rotation.

Rest
The difference between a continuously grazed exercise yard and a productive MIG pasture is rest. Just like a hay field, a pasture grows and yields best when harvest occurs all at once and then is allowed to rest, regrow and restore root reserves before the next harvest. In MIG this is accomplished by dividing the pasture and rotating the herd through divisions or paddocks so that at any one time most of the pasture is being rested. Longer rest periods are gained by dividing the pasture into more paddocks. For example, a farmer might have 30 cows on 30 acres of pasture. At one extreme, he could set stock with a single, large paddock, providing no rest for the pasture. By the end of the first month, it will look like a pool table and he will be feeding hay.

Dividing the pasture into 4 paddocks would result in one week's occupancy with 3 weeks of rest for each paddock. This change can easily double the yield of cool-season pastures. Dividing this pasture into 30 one-acre paddocks would provide for a one-day grazing event with 29 days rest for each paddock. Over the course of a season, depending on weather conditions, a well-managed cool-season pasture can yield 6 to 10 tons of forage dry matter per acre. There is a finite number of growing days during any season. The more days of rest a farmer gives a pasture during that season, the better the forage yield and quality will be.

Many producers use an incremental approach to move into a MIG system, using portable electric fencing to gradually divide pastures into smaller paddocks. A couple of years of experimenting usually reveals an efficient paddock layout. Later, more permanent fencing, lanes, and watering equipment can be installed when the farmer is confident the system will work.

Intake
A cow's day is roughly divided into 8 hours of grazing, 8 hours of ruminating, and 8 hours of sleeping. Her total intake for a day is limited by the number of bites she is able to take during those eight hours of grazing, so the size of each bite makes a big difference. If a pasture is tall, dense, and relatively uniform, each bite will be big and intake will be maximized. If the pasture is thin and sparse, the animal will have to take more bites, use more energy doing so, and may still not get full. The result is reduced production. Although many dairy farmers worry about forage quality when switching to a grass-based system, the most limiting factor in milk production on pasture is intake.

There is no magic to making a pasture tall, dense and uniform. It takes good management and time. Promoting a diverse biological community enhances mineral cycling, water use, and energy capture. Productivity improves as the system matures.

Residual
What is left behind in the paddock is as important as what the animals harvest. A grass plant must re-grow following each of the six to eight grazing cycles of a typical growing season. Leaving adequate green residual will allow grass plants to re-grow without relying too heavily on root reserves to generate new photosynthetic tissue.

In general, the volume of top growth is reflected by a similar volume of root growth. Removal must be balanced so that enough leaf tissue remains to avoid depleting root reserves. A good rule-of-thumb is that grazing should "take-half/leave-half" of the sward. What constitutes "half" of a pasture sward depends on pasture density and the species present. Determining how much is "half" is a judgment that needs to be learned through observation. Watch the speed with which paddocks recover after grazing. If recovery is slow in spite of favorable conditions, grazing has been too severe. There are no recipes for this, it must simply be learned for a specific farm, soil type and climate. It will vary within a season and from year to year.

Rotation

During a typical growing season, grass plants allowed to rest will re-grow within about 30 days. In much of the dairy production region of North America, rate of growth varies with temperature and moisture over the course of the growing season. In spring, a much faster cycle of 14 to 21 days is common. In late summer, the cycle lengthens to 45 days or more.

Paddock system in pasture

Determining when a pasture is ready to graze is a matter of balancing yield and quality. Fresh pasture tends to be higher in protein and lower in fiber than dry hay. New, green leaves have the highest quality; stems and seed stalks are lowest. As the plant grows, leaves age, stems develop and quality declines. For cool season grasses, a good balance between quality and tonnage often occurs when the sward is 8 to 12 inches in height.

The smaller the paddock and the more frequent the moves, the more uniform forage quality and yield will be. Many dairy graziers move their herd to a new paddock after each milking, using a stocking density of around 25 cows per acre. Some graziers use a leader-follower system to optimize utilization of the pasture, allowing the milking herd with its high nutritional need to 'top' the pasture, taking the best quality forage. Heifers and dry cows can follow behind the milking herd, cleaning up the remaining lower quality forage.

How much pasture does a cow need? Many graziers get good enough at estimating pasture forage availability that they can do it by eye. For those just starting out, several rules-of-thumb can help with determining paddock size. In general a ruminant will consume about 2.5 percent of its body weight per day in dry matter. For a 1000-pound Jersey cow, that would be about 25 pounds or the equivalent of half of a small square bale of hay.

We can also estimate pasture tonnage, but there is a great deal more guesswork involved in this calculation. In a dense, leafy, vegetative pasture, there could be anywhere from 100 to 300 pounds of dry matter per inch of height. If we assume we have 200 pounds per inch in a one-acre paddock, 8 inches in height, we have about 800 pounds of forage available (assuming we wish to leave half). If we need 25 pounds of forage per animal per day, our estimate tells us that we can feed 32 animals for one day or 16 animals for two days.

There are a number of tools available to help a farmer gain skills in estimating forage availability. These include using a pasture probe or pasture plate or actually clipping and weighing a standard sized patch of pasture. The pasture probe is an electronic device which uses reflectance to estimate sward density. The pasture plate is a simpler device that works by pressing the forage down with a piece of plexiglass or wood and doing a calculation based on the compressed height of the sward. Both need to be calibrated for the particular types of grasses in one's pasture. A more accurate, though time consuming method is to actually clip, dry and weigh standard sized pasture forage samples.

In the long run, the best way to determine paddock size and length of grazing period is to experiment and observe. Beginning graziers should start out giving the herd more than they feel the cows need. The herd can always be left on a paddock for a longer period of time, or a clean-up herd can be brought in. It's far better to leave a little behind than for the animals to run out of forage.

Use of a pasture measuring stick helps judge pasture condition.

Managing pasture growth cycles

As their name implies, cool-season grasses grow best during cool, moist weather conditions (59°F to 86°F). In the moderate climates of New Zealand and the British Isles, growth of these grasses tends to be fairly consistent throughout the growing season. The growth curve in these regions resembles a dromedary (one-humped) camel.

In the temperate regions of North America, where hot, dry conditions occur in mid-summer, the growth curve of these grasses resembles a Bactrian (two-humped) camel. These pastures are characterized by a flush of growth in spring and early summer and another, smaller growth period when weather cools down in late summer and fall. The mid-summer slow-growth period is often referred to as a summer slump. Depending on the season, the slump can turn into a prolonged dormant period for some grass species or in some years it might barely be noticeable.

The more accentuated the growth curve in a farm's area, the more important it is to develop strategies to even out the high and low growth periods. As a rule of thumb, a typical dairy cow requires the equivalent of 1.5 to 2 acres of pasture per year to meet her forage requirements. If pasture plants grew uniformly 12 months of the year, it would be easy to manage a grazing system. Because pasture growth varies not only between summer and winter, but also within the growing season, it is necessary to utilize a number of tools to manage forage production.

Hay curing

The most commonly used strategy is cutting paddocks for hay during the spring flush. Up to a third of paddocks can be set aside for hay cutting. The sward is allowed to grow for four to five weeks, similar to an alfalfa crop, and is mowed, cured and baled for use in winter. Haylage can also be made. These paddocks are then worked into the grazing rotation later in the season when growth has slowed. One or two cuttings can be made, depending on growing conditions.

Some producers maintain dedicated paddocks for hay production, while others rotate which paddocks are set aside for hay cutting each year. For example, a producer might want to set aside the farthest paddocks from the milking parlor for hay production. These paddocks could be seeded to species such as smooth brome, orchardgrass and alfalfa, which perform well under hay cutting. Later in the season, the paddocks can be incorporated into the grazing rotation as they are needed.

Alternatively, the argument can be made that rotating which paddocks are cut for hay can provide management and timing flexibility. Varying harvest methods can also improve pasture species diversity and density and provide an effective means of addressing weed problems or other management issues.

Stockpiling

Stockpiling pasture for later grazing is another means of improving pasture allocation. Fall stockpiling is a common practice involving setting aside up to a third of pasture acres in late summer to be grazed after freezing temperatures have stopped growth. In the Upper Midwest, paddocks are usually set aside in mid-August. For optimal forage accumulation, adequate nitrogen and moisture must be available during the 8 to 10 weeks before the end of the growing season. The stockpiled forage can yield approximately one ton per acre and retains quality fairly well for several months. Many graziers successfully extend the grazing season through early winter with stockpiled pasture.

While a few graziers outwinter their milking herds, it is more common for stockpiled forage to be utilized for non-lactating animals with lower nutritional requirements. Cattle can graze effectively through up to 8 to 10 inches of snow cover, but as the foraging becomes more difficult and temperatures become colder, energy use can exceed the energy contained in the forage. Rain, snow, and temperature fluctuations reduce forage quality of stockpiled forage and make stockpiling for spring use less effective.

Stockpiling will also set up a staging of paddock growth in the spring. Paddocks grazed right up to frost go into the winter with a minimum of root reserves and tend to green up more slowly in spring. Stockpiled paddocks allowed to grow through the fall go into winter with root reserves intact if they are grazed after growth has stopped. They tend to green up and be ready to graze earlier in spring.

Rotation

Although rarely used, summer stockpiling can be an effective tool to help manage variability in pasture growth, especially if late summer drought is expected. For summer stockpiling, the paddocks are grazed into early summer, then clipped to remove seed stalks and mature material. For most cool-season grasses, once they are past the reproductive period, they should put on higher quality, leafy growth for the rest of the season. These paddocks are then set aside to accumulate tonnage to be grazed during the hot, dry months. Selecting paddocks with a high legume component will further improve quality and tonnage.

Pasture fertility and balancing nutrients

For many organic farmers it is valuable to think about fertility on a whole farm basis. Ideally, a farm's cropping system is devised to minimize the export and import of nutrients and to maximize the effectiveness of mineral cycling within the production system. The harvest of meat and milk from pasture-raised animals represents a relatively small removal of nutrients from the farm. If a healthy mineral cycle exists, little additional fertility will be needed for pastures to perform well.

The pasture ecosystem relies on the microbial digestion of plant material by ruminants to contribute to the cycling of nutrients. Without them, decomposition by free-living microbes would slow or cease during winter and dry periods. With good management, a pasture system improves over time. Soil organic matter increases, creating greater water holding capacity, increased mineral availability and greater efficiency of energy capture. The result is better intake, higher milk production and more profitability.

Monitoring organic matter, soil pH and macro-nutrient levels in the soil through soil testing should be a part of a grazier's management program. A three-year interval is usually adequate, with a composite soil sample taken for every 5 to 10 acres. Areas with significantly different soil types or plant growth patterns should be sampled separately. Because a pasture is a community of several plant species, it is most useful to think in terms of optimum ranges for the primary nutrients.

For phosphorus, optimum fertility levels for cool season pasture are 11 to 23 parts per million (ppm) for silt loams and 23 to 32 ppm for sandy soils. For potassium, target soil concentrations are 66 to 120 ppm for silt loams and 61 to 90 ppm for sands. The optimal pH range is 6.0 to 7.0.

The role of supplementation

Most grass-based dairy producers supplement pasture forage with corn or a more complex mixed feed ration, as well as minerals. It is not necessary to change a supplementation program when converting to a pasture forage system except to make sure that the ration remains balanced.

If supplemental grain or silage is produced on-farm, it is easy to maintain a balanced nutrient cycle, with manure going back onto the crop ground from whence the feed came. However, many graziers minimize production costs by dispensing with crop production and buying in grain. Nearly all dairy producers rely on purchased minerals. Be aware that you are bringing nutrients onto the farm when you do this. Such purchased ration supplements can provide a means of augmenting pasture fertility or exacerbate an oversupply or imbalance of soil nutrients.

Nitrogen

The biggest fertility challenge for the organic pasture producer is managing nitrogen. Cool season grasses respond dramatically to nitrogen fertilizer, but there is no economical organic source to provide the 50 or so pounds of nitrogen per grazing cycle needed for maximum growth (The University of Wisconsin Extension recommendation is 150 pounds per acre per year in a split application). Most organic producers rely on legumes and manure for nitrogen sources and think in terms of optimizing yields rather than maximizing them.

In fact, nitrogen is abundant in most mineral soils with moderate organic matter levels. The problem is that it exists in forms that are not always readily available to fully meet the needs of rapidly growing grass plants. The most important thing you can do to make nitrogen available is to enhance mineral cycling by encouraging a healthy biological community in the soil. Balanced levels of carbon and other minerals will allow rapid breakdown of organic matter and release of nitrogen for plant growth.

Legumes

Legumes are a key element in the pasture plant community for three reasons. The first is their contribution of biologically fixed nitrogen through root associated *Rhizobium* bacteria. This nitrogen is primarily made available as the top growth of the plant is consumed and excreted by cattle. In addition, some of the nitrogen fixed by legumes becomes available as root tissue ages and breaks down.

Legumes, such as clover, enhance pasture quality and palatability.

Legumes also enhance pasture forage quality and palatability. Because they are more tolerant of hot, dry weather, legumes can play an important role in providing pasture tonnage in late summer. Most people strive to maintain approximately 50 percent legumes in their pastures. The most commonly used legume in Upper Midwestern pastures is red clover, an inexpensive, short lived perennial. Ladino clover is gaining in popularity because it yields similarly and is more persistent, although at the current time the seed is more expensive.

Alfalfa is not as popular for pasture use and some graziers observe that their cattle find it less palatable than the clovers for fresh grazing. It is commonly used, however, in situations where paddocks are dual use, for hay and grazing. It is also the best adapted legume for sandy soils and drought conditions. Because of its deep taproot, alfalfa can improve soil nutrient use by drawing nutrients up from lower soil horizons.

Other, less common legumes include alsike clover, which tolerates low pH and wet conditions, and birdsfoot trefoil which requires a long rest period and is not bloat causing. Sweet clovers, sometimes used as green manure crops in organic systems, are not desirable in a pasture setting. These biennial clovers (*Melilotus* species) produce an anti-quality compound, coumarin, which gives its tissues a bitter taste and can cause health problems in livestock.

Legumes are regularly frost-seeded into existing pastures to maintain adequate sward composition. Frost-seeding is done in late winter when the top few inches of soil are freezing and thawing in response to the warmth of the sun. In late afternoon or early morning, the ground is firm enough to drive over to broadcast the seed on the surface. By mid-day, the ground is warmed and soft. The freezing and thawing allows the seed to be worked into the soil so that it is ready to germinate when temperatures moderate.

Frost-seeding 1 - 2 pounds of ladino or 2 - 3 pounds of red clover per acre every 2 - 3 years is an inexpensive way to improve pasture yield and quality and to maintain nitrogen fixation.

Utilizing manure resources

Although the bulk of manure in a pasture based system is "self-spread," some manure spreading on pastures may be needed. Manure collected from the milking parlor or barn is a valuable resource for both crop land and pasture. If crop land exists on the same farm, it is likely that manure will be needed to spread on these acres. If not, or if it is determined that a pasture area would benefit from additional manure application, composted manure is preferable.

Manure will be distributed naturally by cows as they graze and move between paddocks.

Some producers have found that there is no refusal when cattle are pastured on paddocks spread with composted manure. If non-composted manure is used, it should be applied immediately after the paddock has been grazed. Depending on weather conditions and the speed with which the manure breaks down, there may be some refusal the next time the herd enters the spread paddock. A healthy, biologically active soil will hasten breakdown and cycling of both spread manure and dung pats.

Manure Distribution on Pasture

Ideally, manure should be distributed as uniformly as possible on the pasture, returning the nutrients to be recycled. Small paddocks and frequent moves improve manure distribution on pasture. Places where manure and urine concentration tend to occur include around watering and feeding areas and where the animals "camp" to ruminate. A single tree in a pasture is a real draw and should be avoided.

Water

Some graziers provide water only at the barn and give their cattle time to drink what they need before and after milking. However, it is generally agreed that providing water on pasture improves milk production. Researchers have found that if cattle are never more than 700 feet from watering stations on pasture, they will drink individually rather than going to water as a herd.[22] As the distance from water increases, drinking becomes more of a group activity and a great deal more manure and urine is deposited around the watering site. So, having watering stations in sight within each pasture improves nutrient distribution. Although many graziers have permanent watering stations, having the flexibility to move these sites around in the pasture will also improve nutrient distribution and reduce the wear and tear on the pasture around the tank.

Weed management

When considering weed control in a pasture system, the first thing to do is revisit the definition of a weed. The perennial nature of the pasture plant community reduces the prevalence of annual, successional weeds. The fact that there are only a few plant species that are unpalatable or toxic further reduces the issue of weed control to a very manageable level in most cases.

For example, while quackgrass can be a serious problem in organic cropping systems, it is one of the highest quality, most productive grasses in a pasture setting. Because the pasture system functions more effectively as biological diversity increases, the presence of a wide variety of grasses, legumes, and broadleaf plants is an asset.

Many broadleaf weeds are high in protein, low in fiber and may contain valuable concentrations of micronutrients. Different plant species utilize different parts of the soil profile uniquely and accumulate a unique blend of minerals from the soil. Their presence can enhance the nutritional quality of a sward.

The unpalatable exceptions include such species as thistles and burdock. There are no commercial herbicides allowed under the National Organic Program available to tackle these weed problems, but there are several tools that can be used. Probably the most important weapon against weeds is, once again, good pasture management. By managing to favor desirable grasses and legumes, a farmer goes a long way toward limiting the capacity of weeds to cause problems in pasture systems.

Clipping is the most widely used means of keeping weeds from going to seed and can be very effective in controlling annual and biennial weeds. Burdock and several thistle species are biennials, which means they are vegetative in their first season and produce flowers and seeds during their second growing season. If clipping is timed when these weeds are in mid- to late-flowering, seed production will be halted and the plant will die. One or more clippings, usually following grazing, may be needed.

---

[22] *Gerrish, Jim, P.R. Peterson and R.E. Morrow. 1995. Distance Cattle Travel to Water Affects Pasture Utilization Rate. Forage Systems Newsletter, Missouri Agricultural Experiment Station Vol 4 #2. http://aes.missouri.edu/fsrc/research/afgc95h2.stm*

One of the biggest problem plants in pasture systems is Canada thistle, a perennial. There is some evidence that clipping actually encourages this spreading plant to send out more roots. If clipping is to be effective it must be timed precisely to remove mature flowers and seed heads. This timing attempts to attack the plant at the point when it has expended a maximum amount of energy on reproduction.

Vinegar is another possible weed management tool for organic farmers. Vinegar is a non-selective desiccant. The small amount of acetic acid present at 5 percent in grocery grade vinegar is enough to burst plant cell walls, and the plant turns brown within 10 minutes of application. Although it is effective on top growth, there is no systemic or residual activity and it is unknown whether it can kill the root system of perennial weeds like Canada thistle. Other negatives include the cost of applying to large areas, its capacity to reduce soil pH if used heavily, and the fact that it is legally not registered as a herbicide, either for organic or conventional use.

Another possible approach to controlling problem weeds is to spot spray with a molasses or salt solution. The idea behind this approach is to make the weeds more palatable to cattle and control them through selective grazing. There is no data to support the effectiveness of this approach, but logic would suggest that it might work if done early in the season when plants are tender.

Some producers may be in a position to consider multi-species grazing. Sheep and goats tend to graze more broadleaf plants, and may be effective at bringing weeds under control. Goats are well known for their capacity for grazing and browsing unpalatable herbaceous and woody plant materials.

Weed control is an on-going effort for pasture managers. Understanding the biology of the weed species present can provide direction for control strategies. The weed seed bank of a farm will dictate to some degree how much of a struggle weed control will be. Other sources of weeds could be seeds blown in from other areas or brought in with purchased hay or feed.

A few words about what to plant

When beginning a pasture management system, one should start out by utilizing any existing pasture that is available and incorporating mature, grassy hayfields into grazing rotations. In many areas there is a viable seed bank of grasses and legumes that will express itself when a system of rotation and rest is applied. This existing plant community has the advantage of being well adapted to the local climate and soil conditions. It's always valuable for farmers to determine what they have to work with before investing any additional money and resources on "new and improved" varieties. Renovation of existing pasture seldom pays off.

However, when one needs to seed down a new pasture, starting with a high quality mix or a custom blend is worth the extra expense. This is not the place to save pennies. It is worthwhile to avoid the inexpensive pre-mixed pasture blends on the market. They often contain low quality varieties, such as endophyte-infected tall fescue or annual ryegrass. In selecting seed, it is wise to determine which species are best adapted to the farm and to work with a company that can custom blend pasture seeding mixes.

Many people think in terms of simple mixtures of one or two grass species and one legume. One might vary the species used on different parts of the farm that may have differing soil types, slopes, aspects or planned uses. For example, one might plant alfalfa and smooth bromegrass on a dry, south facing slope, whereas a high fertility soil with plentiful moisture holding capacity might be a good place for perennial ryegrass. This approach makes management easier and provides diversity on a farm-wide basis.

Another approach is to consider a more complex mixture of three to five grasses and three to five legumes. This approach would have a different set of advantages: it would enhance the biological diversity within each pasture and allow the best-adapted species to express themselves across the landscape.

Whichever approach is taken, it is worthwhile to start the process by soil testing and evaluating weed pressure. Seeding down a new pasture on former crop land presents an opportunity to get weed problems under control and to correct any soil fertility or pH imbalances that might exist.

Conclusions

Management intensive grazing can be a very low-input system and is well suited to organic dairying. Like many aspects of organic production, there is no template to follow. The most important things a farmer needs to know about managing a pasture-based dairy system can't be learned by reading a book. Thus, the focus of this chapter has been on understanding the basics of how pasture systems function. How one uses this understanding in day-to-day decision making is the single most important factor influencing productivity. The best managers are those who never stop observing, learning, and adapting.

Farmer Profile: Francis Thicke - Fairfield, IA
Paul Bransky

Radiance Dairy, owned by Francis and Susan Thicke, is a 65-cow organic dairy farm in southeast Iowa, succeeding with grass-based Jersey milk and local customer connections. "I sell all my products locally; I know where it goes," Francis explains, "so I feel a responsibility to do the best that I can."

For Francis, "the best" means producing high CLA (conjugated linoleic acid) milk from grass while improving soil. "What I like to see is organic as an ecological system, where you use rotations to help provide fertility and pest control. Sometimes we now see organic going to industrial agriculture, where instead of using conventional inputs, we use natural inputs from the approved list. We don't change the whole system. If you look at natural ecosystems, like the prairie that was here 150 years ago, it doesn't have pollution or waste. The waste from one species becomes the food for another species. I'm trying to see my farm as an ecology, to see how one piece can play off the other pieces and how I can simplify things as much as possible. That's the goal of organic, in my mind."

Francis Thicke

Francis' entire 236-acre farm is seeded down to pasture, divided into about sixty paddocks of two to three acres, plus large areas mostly for hay production. "I think grazing is central to an organic dairy farm, because it really makes an ecological system," he says. "I'm going towards the goal of feeding 100 percent grass." Francis currently feeds eight pounds of grain per day. "Cows get a fresh stretch of pasture after each milking. Ecologically it's a much better design to have the cows harvest their own feed and spread their own manure. Of course, milk production may not be as much, but energy use and profitability are better."

Francis sees better grass species diversity with rotational grazing compared to continuous grazing. "We try to keep our pasture mix as diverse as we can," he says. Brome is their basic grass, along with timothy and low alkaloid reed canary grass. Francis frost seeds clover in the spring, but doesn't normally seed grasses. They started out with red clover as their main clover, but switched to Alsike white clover. "I like it because it seems to last and it spreads with stolons." He also likes Kura clover. "Kura gets really thick. If it didn't start so slow it would be a lot nicer." Francis does not like perennial rye grass. "It's too high input of a system," he says. "It doesn't persist, so you have to keep replanting."

Chapter 11 Pasture Management

Francis says that his cows will not eat fescue. He tries to control it by overwintering them on the pastures where it grows, then rotavating and planting an annual the following spring. He hopes that after two years of this treatment the fescue's seed bank will be completed, and he can establish perennial pasture again.

"We're trying to extend the grazing season as much as we can," Francis notes. In 2003 they started grazing on March 31. Located in southern Iowa, "we start when the grass is just taking off."

"Two mistakes I think people make when they go grazing: One, they don't start early enough and they don't graze aggressively enough early, and, two, they are too aggressive late in the season and don't give the grass enough rest. We start out early so we can get the farm sequenced. I also think the grass tillers (spreads underground) a little more if you graze it a little bit early on."

Early in the season Francis rests his paddocks ten to twenty days between grazings. Later, in July and August, he extends to thirty to forty days in a normal year, up to fifty days during a dry year. "I believe strongly in rest periods," he says. "I think that builds soil and plant density, and it is also better for the animals to have a longer rest period, where the plants are a little more mature. You have taller grass with a deeper root; when it is grazed off, you're sloughing more organic matter deeper into the soil."

Francis also feels that taller grass is better for his cows. "If you look at young, lush grass, it's 20-25 percent protein — like a protein supplement. Look at what comes out of a cow's back end. It's not good with lush grass. I think more fiber is needed. Maybe you aren't going to get as much production, but I think it's more sustainable in the long run."

Recently Francis started wintering cows outside in different paddocks each year, following a pattern his brother Art has used successfully. He says that wintering cows on paddocks saves a lot of manure handling in the spring. He puts round bales in rows then advances the fence wire to limit feeding to one row at a time. His cows lay on residual hay from each previous bale, so the manure tends to spread evenly over the pasture throughout the winter. "I kind of learned how far to space the bales out to get the manure out in the pattern I am looking for." He says Art's winter paddocks are noticeably more productive for several years afterwards. He also likes moving just wires all winter, instead of tractors and hay.

Biodiversity
Harriet Behar and Jo Ann Baumgartner (Wild Farm Alliance)

Under the National Organic Program, the definition of organic production includes a reference to conserving biodiversity: Subpart A – Definition 205.2 Organic Production. A production system that is managed in accordance with the Act and regulations to respond to site-specific conditions by integrating cultural, biological, and mechanical practices that foster cycling of resources, promote ecological balance, and conserve biodiversity.

What does this mean to the organic dairy farmer and what benefits do they gain by incorporating a plan for biological diversity into their organic system? Biodiversity includes the variety of all forms of life, from bacteria and fungi to grasses, ferns, trees, birds, reptiles, mammals and insects. This also encompasses the diversity found among all levels of organization, from genetic differences between individuals and populations (groups of related individuals) to the types of natural communities (groups of interacting species) found in a particular area. Other processes such as predation, symbiosis and natural succession are also part of biodiversity.

Of the 200,000 plants and animals now known to exist in the U.S., fully one-third are at risk, with 400 species already lost to extinction and another 100 missing. Agricultural lands comprise roughly two-thirds of the continental U.S. and all farmers have a great opportunity to aid in conserving the biodiversity on the lands they manage. When balancing the changes agriculture brings to the land and determining what biodiversity should be conserved, all things are not equal, and should not be given equal weight. In other words, the loss of an eagle or a wetland is not balanced by the gain of a pigeon or a hayfield. However, managing a farm with environmental health in mind has both agricultural and ecological benefits. Some of the many ways in which farmers promote ecological balance are by:

- Encouraging biodiversity, which can help the farmer take advantage of nature's ecosystem services, such as pollination, pest control, beneficial predation and fire, flood and erosion control, nutrient cycling and improved water quality and quantity.
- Thinking about the farm as a whole ecosystem; maintaining and restoring linkages, connectivity and wildlife corridors to strengthen regional networks of conservation areas.
- Planning with the prevention of the introduction and spread of non-native,

Hedgerows and buffers add diversity and encourage pollinators.

invasive species in mind. The spread of invasive plants, animals and pathogens is the second major threat to biodiversity in the U.S., with the cost an estimated $138 billion per year. The farmer can participate in state, federal or non-governmental habitat conservation and invasive species control programs.

- Planting crop rotations so that some fields always provide food-- either intentionally planted wildlife food crops, or crop leftovers. In-field pollinator and natural enemy insectary plants can be grown to provide cover and habitat for beneficial insects and other wildlife.
- Managing the farm to accommodate sensitive life stages, such as nesting, by delaying hay and grain harvests enough to allow ground-nesting birds to fledge. Or if non-crop vegetation is to be cleared, it may be done before bird breeding season so birds do not become established. Alternate mowing of field grasses and cleaning alternate sides of ditch banks provides wildlife refuges in undisturbed sites and allows for filtering processes important for water quality and the needs of aquatic species.
- Controlling pest species in ways that do not have negative effects on desirable native species.
- Designing, installing and maintaining sequentially flowering hedgerows and windbreaks to benefit priority and other native species. These can be planted in wide swaths where appropriate and have a layered complexity of native plants that supports a diversity of wildlife including beetles, pollinators, natural enemy insects and birds. Hedgerows and windbreaks can be linked to natural areas on and off the farm, where feasible.
- Planting or conserving native trees (even as snags) in field margins as roosting and nesting habitat for birds and other beneficial animals.
- Installing boxes to support insectivorous birds and bats as well as predatory birds.
- Integrating barn cats humanely and responsibly to reduce pressure on birds and bats.
- Preserving wide riparian areas in pastures that benefit native species and may serve as a wild-way or link to neighboring habitats and contribute to broader conservation network. In erosion prone areas, runoff may be collected in tail-water ponds, instead of eroding off the farm. The farmer can give extra attention to protect special habitats. Restoring damaged areas, if needed, with native perennial grasses and forbs for the benefit of livestock and wildlife.
- Managing the frequency, intensity and timing of livestock grazing to minimize negative impacts to soil, grazed areas and ecosystem health.
- And lastly, avoiding the use of guns, traps, or poison to protect livestock from native predators, but instead using a variety of practices, for example: Guard animals, such as llamas, donkeys, or dogs, to help protect livestock. Pasture use can also be scheduled for when predation pressure is low.

Chapter 12

Understanding Organic Certification

Now that you have a basic grasp of how to produce milk organically, it is time to look at the process of organic certification so that your product can be sold into the organic marketplace. The following is a comprehensive explanation of the US Department of Agriculture's "National Organic Program: Final Rule" which was put into the Federal Register on December 21, 2000 and implemented in October of 2002.

Understanding Organic Certification
Joyce Ford and Lisa McCrory

Organic certification is a process where an impartial third party, or certifying agent, reviews and evaluates a specific operation, such as a farm, ranch, or processing facility, to determine that the operation meets organic standards. Certification assures consumers that the product they are buying is, indeed, "organic." Certification is required when marketing organic products wholesale. Organic products, including dairy, may be processed several times and added as ingredients to other organic products before the final products are sold to consumers.

The Organic Foods Production Act (OFPA), implemented in October 2002, requires that anyone selling a product labeled as "organic" must be certified by a certifying agent that has been accredited by the US Department of Agriculture (USDA). The only organic farmers and handlers exempted from the requirement of certification are those selling less than $5,000 of organic agricultural products annually. Even those farmers need to comply with National Organic Program (NOP) requirements, including keeping appropriate records. Products from "exempt" operations must be sold directly to consumers. They cannot be used as ingredients in processed products without being certified.

The certification process

If considering organic certification, contact a certifying agent well in advance to understand specific requirements and begin using organic management practices. Some agencies work closely with an accompanying non-profit association to guide farmers through the transition process, but all will be able to provide certification standards, Organic Farm/Livestock Plan applications, and other helpful materials. The transition process can take from one to over three years, depending on the changes that need to be made on the dairy farm.

To qualify for certification, prohibited substances must not have been used on the land for 36 months prior to harvest of an organic crop. Prohibited substances include synthetic fertilizers, herbicides, insecticides, sewage sludge, and genetically engineered seeds, inoculants and other products. This 36-month period is the transition[23] period. All fields may not qualify at the same time. For instance, hay fields or pastures may not have had prohibited substances used for several years and may qualify immediately while corn acres may need the 36-month transition period. In that case, non-organic crops can be sold from these transitional fields as long as adequate records are kept to verify sales.[24]

Dairy cows must be managed organically for one year prior to certification, including feed, health care, living conditions and record keeping. Until June 2006, dairy farmers converting "whole herds" to organic production have the option of feeding for nine months at least 80% certified organic feed or feed raised on land included in the organic farm plan and managed organically, followed by 100% certified organic feed for 3 months.

In October 2005, Congress amended the Organic Foods Production Act to allow dairy farmers to feed farm-raised "3rd year transitional feed" for one year prior to the production of certified organic milk. The change to OFPA needs to be fully implemented through changes to the regulation, but it is anticipated that the allowance for one year of third-year transitional feed will replace the "80/20" provision that was struck down by the First Circuit Court of Appeals.[25]

The following outlines the steps included in the certification process. The dairy farmer must establish and implement an Organic System Plan, which provides detailed information on land resources, crop production, livestock management, and inputs used. Before certification is granted, the dairy farmer must have an on-site inspection, be approved by the certifying agent, and comply with

---

[23] *During the 36 month transition period, fields and crops are managed organically. These fields are "in transition" to certified organic production.*
[24] *It should be noted that the farmer needs to keep records for non-organic crops similar to organic crops, and maintain separate storage areas for non-organic crops.*
[25] *The use of 20% conventional feed during the first 9 months of transition was allowed under the NOP regulations. However, the Circuit Court ruled that the Organic Foods Production Act does not allow the use of nonorganic feed during transition. Rule revisions are being developed by the USDA to bring the regulations fully into line with Congress' amendments to OFPA and the court's ruling.*

the NOP. Unless the organic dairy farmer is not growing any crops and is buying all feed, the farmer must develop and implement both Organic Farm and Livestock Plans.

Step 1: A certifying agent is chosen and an Organic Farm and Livestock Plan information packet is requested by the farmer. An Organic Farm/Livestock Plan includes the following:

- A description of farm practices and procedures to be performed and maintained.
- A list of all materials planned to be used as production inputs, including seeds, fertilizers, silage inoculants, feed supplements, livestock health care products, dairy barn detergents, sanitizers and teat dips.
- A description of the monitoring practices and procedures to be performed and maintained. (Monitoring may include soil tests, visual observations and individual cow health care records.)
- A description of the record-keeping system.
- A description of the management practices and physical barriers established to prevent commingling of organic and non-organic products and to prevent contact of organic production and handling operations and products with prohibited substances. Field buffers, equipment cleaning and separate storage areas are examples of this type of information.
- Additional information as deemed necessary by the certifying agent. Examples are farm maps, water tests and 3-year field history sheets.

The certifying agent asks questions to assess basic eligibility and determine if additional applications or questionnaires (i.e., processing) are appropriate. The farmer is sent an application packet that usually contains the NOP Final Rule, certifying agent procedures, Organic Farm Plan and Organic Livestock Plan applications, examples of records and other resources and detailed instructions. There is generally a fee for this packet.

Step 2: The Organic Farm Plan and Livestock Plan applications are completed by the dairy farmer, including farm maps and 3 year field histories for all fields, including pastures. In the Organic Farm Plan application, strategies for improving soil health, crop rotation, and the control of weeds, insects and crop diseases are described. All inputs, including seeds and seedlings, fertilizers, and weed, pest, and disease control products, are listed. Label information and input use records are kept on file by the

Approved materials may have an OMRI label that shows the item is approved for organic production.

farmer. All equipment and crop storage areas are listed, and crop harvest plans are detailed, including how the integrity of organic crops is protected during production, harvest, transport and storage.

Questions on sources of cows and replacement stock, feed and feed supplements, housing, manure management, access to outdoors and/or pastures, health care plans and products used and milkhouse and pipeline system cleaning products are answered in a Organic Livestock Plan application. Somatic cell count information may be requested by the certifying agent. Labels for all purchased feeds, feed supplements, health care products and milkhouse cleansers and sanitizers should be kept. If the product is not clearly approved for use, the labels may need to be sent to the certifying agent for review. All input products should be approved by the certifying agent prior to use.

Marketing strategies are also described in the Plan. If use of "organic" labels is planned, sample labels must be submitted. The completed Organic Farm Plan and Livestock Plan applications and attachments, licensing agreements and fees are returned to the certifying agent.

Step 3: The certifying agent reviews both questionnaires, and any submitted records, letters, labels and other documentation for completeness and to determine if the applicant has the ability to comply with the NOP.

Step 4: If the applicant appears to have the ability to comply, the certifying agent assigns an organic inspector. It is the inspector's job to verify information and assess compliance, but he/she does not make the certification decision. The inspector sets up an appointment with the farmer. Depending on the complexity of the operation, the inspection may take 3-8 hours. First time inspections, inspections with multiple locations and inspections where the farmer is not well organized take longer.

The inspector reviews all aspects of the Organic Farm and Livestock Plans with the farmer. All animals, crops and fields, including adjoining land uses, buildings and equipment involved in the operation are inspected and evaluated for contamination risk and compliance. Records and labels are reviewed to insure monitoring and compliance. The inspector completes an affidavit with the farmer and reviews findings, additional information requested and issues of concern in an exit interview. An inspection report with supporting records and label information is sent to the certifying agent.

Step 5: The entire file is reviewed by a certifying agent official or review committee. More information may be requested before a certification decision is made.

Step 6: Certification is granted if the certifying agent determines that the operation complies with the NOP. A certificate of organic operation is issued. Once certified, an operation's organic certification continues until withdrawn by the producer, suspended or revoked. Upon certification, the producer may sell certified organic products. Organic practices must be followed throughout the year, including records that track organic products from production through harvest, storage, transport and sales. The USDA Organic Seal may be used on certified organic products. A certification update, including on-site inspection, is required annually.

Noncompliance issues may be identified during the review of the inspection report and supporting documentation. Certification may include requirements for correction of minor noncompliances within a specific time period as a condition of certification. Examples of minor noncompliances are non-implementation of crop storage records or individual cow health records, inadequate maintenance of field buffers or the need to manage manure better in order to minimize water contamination. A good understanding of organic standards and preparation can reduce the number of noncompliances. Certification can be denied, suspended or revoked if the applicant is not willing or able to comply with the NOP requirements.

Choosing a certifying agent

As of late 2005, 99 international and US based organizations have been accredited by the USDA to certify farmers (both crop and livestock) and/or processors under the National Organic Program (NOP). All accredited certifiers enforce the NOP, and most are accredited for certification of crops, wild crops, livestock and/or processing. Not all certifiers are accredited to certify organic livestock, however. Not all accredited certifying agents operate in all regions, although most regions offer a choice of certifying agents. For a complete list of USDA accredited certifying agents, refer to the NOP website (http://www.ams.usda.gov/nop/CertifyingAgents/Accredited.html). Be sure the certifying agent you choose is accredited to certify organic farms and livestock operations and operates in the region you live.

In choosing a certifying agent, marketing needs, differing costs of certification and the types of services provided by each certifying agent should be taken into consideration. Some certifying agents are connected to a non-profit educational organization[26] that sponsors conferences, workshops and field days and publishes educational materials. These non-profit organizations are great sources of information for transitioning farmers. Neighboring organic dairy farmers can offer input on the certifying agents they are using.

If you are planning to sell your milk to an organic milk processor, contact them for information on specific certifying agents they may require or prefer. Crops or livestock planned to be marketed in Europe need to be certified by

---

[26] *Examples are Northeast Organic Farmers Association-Vermont (educational non-profit) and Vermont Organic Farmers (certifying agent) or Oregon Tilth, Inc. (educational non-profit) which administers Oregon Tilth Certified Organic (certifying agent). MOSES is the educational non-profit for Midwestern-based certifying agents, including Midwest Organic Services Association (MOSA) and the Midwestern chapters of the Organic Crop Improvement Association (OCIA).*

an ISO Guide 65 approved or IFOAM-accredited certifying agent.[27] Types of services offered by certifying agents vary as well as costs. Additional services offered may include certification for markets other than the U.S. (Europe and Japan), certification of transitional crops (not regulated by the NOP), issuance of transaction certificates per load of organic product sold, issuance of export documents and review of inputs for approval. For more information on certifying agents, visit The New Farm Guide to US Organic Certifiers at www.newfarm.org/ocdbt/.

Fees for certification will vary with the certifying agent. Fees generally include: a fee for the application packet, general certification fees for crops, livestock and/or processing and inspection fees. Some certifying agents charge a flat fee while others have a sliding scale based on acreage or projected amount of organic sales. Some have minimal upfront fees and charge an additional % based on actual organic sales. Most certifying agents charge separate fees for certification of organic crops (even when used for feed) and livestock. There may be separate fees for first time certification, out of state certification, or on-farm or contract processing certification. Some agencies have application deadlines and charge late fees. Some have additional charges for specific services, such as export certificates or review of products. Fees are paid annually. A survey of several agencies' fee schedules showed certification fees for a dairy farm grossing $80,000 in the range of $1,100 to $1,272, while a dairy farm grossing $250,000 might pay $1,300 to $2,800 per year. Inspection fees can also be a separate expense, covering the inspector's time and travel expenses. Certifying agents generally try to group inspections to minimize costs.

Dairy producers not processing their own dairy products are encouraged to contact a processor at the same time that they contact a certifying agent to

> **Most states administer a federal cost-share program to reimburse producers and handlers for 75% of the costs of certification, up to $500 per year. Check with your State Department of Agriculture for more information.**

ensure they have a market for their organic dairy products. The processor may need time to change the milk pick-up route or may not be interested in a producer's milk if they are not located in an area where milk is already getting picked up. New milk is most often added in late summer, fall or winter. Processors often have strict milk quality and animal housing standards that exceed the NOP, and may choose not to work with a producer if they are not up to their standards. An important marketing step when the farmer is in the last year of transition is obtaining a letter of commitment from a processor indicating the date of first pick-up, the pay price, etc.

---

[27] *ISO is International Organization for Standardization. International Federation of Organic Agricultural Movements (IFOAM) is an international non-governmental organization. For more information on international certification requirements, refer to the IFOAM website, www.ifoam.org.*

Overview of practices and materials allowed in organic dairy farming

In the United States, organic management practices and allowed materials have evolved over the last 50 years, culminating in the National Organic Program (NOP). This farming system is not "just what a farmer is <u>not</u> applying to fields and crops." Organic farmers strive to improve soil and plant health, cycle resources, promote ecological balance and conserve biodiversity using cultural, biological and mechanical practices. The NOP outlines allowed practices and materials as well as specific requirements for certification. In the following discussion, the appropriate NOP sections are cited in parentheses in order to refer to the original wording in the National Organic Program Final Rule, 7 CFR Part 205.

Organic Crop Management Practices and Approved Materials
A fertility management plan must be implemented that maintains or improves the physical, chemical and biological condition of the soil and minimizes soil erosion (NOP §205.203). Strategies may include (but are not limited to) diverse crop rotations, cover crops, green manure crops, use of purchased soil amendments or micronutrients, compost and/or manure, periodic soil/plant tissue tests, incorporation of crop residues, subsoiling and soil inoculants. The crop rotation plan must maintain or improve the soil, provide pest management, manage deficient or excess plant nutrients and provide erosion control (NOP §205.205).

If the use of fertilizers or other inputs is planned, care must be taken to ensure the inputs are approved for organic production. In general, naturally mined minerals, such as limestone, rock phosphate and potassium sulfate, and micronutrients, such as sulfur, boron and copper, are approved soil amendments. Synthetic soil amendments are prohibited unless they are specifically listed on the NOP National List (NOP §205.601). Natural (or nonsynthetic) materials that are prohibited are listed in NOP §205.602. Keep in mind that synthetic products can be added to the NOP National List upon the recommendation of the National Organic Standards Board (NOSB) to the Secretary of Agriculture and published in the Federal Register as an amendment to the NOP. Until products recommended by the NOSB are published in the Federal Register, they are prohibited by law. The Organic Materials Review Institute (OMRI) reviews brand name products and ingredients for approval based on the NOP National List. OMRI semi-annually publishes lists of generic materials and brand name products that are approved for use on organic crop, livestock and processing operations. Many certifiers also review both generic and brand name inputs. Check with your certifier before purchasing and applying any inputs to make sure that you follow the agency's policies and procedures.

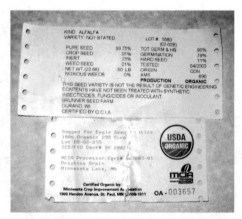

(Bottom tag) Seeds must be certified organic.

It is the responsibility of the farmer to be sure all inputs used are approved prior to use. The certifying agent can confirm whether a particular input is approved. Soil tests may be required, especially if micronutrient soil amendments are applied.

Manure must be managed to maintain or improve soil organic matter content and so that it does not contribute to crop, soil or water contamination. In cold climates, winter application of manure may be prohibited or regulated, depending on the certifying agent. Manure does not have to come from organic livestock, but it cannot contain residues, such as heavy metals, that could contaminate crops, soil or water. (NOP §205.203(d) and §205.239(c))

Seeds must be organically produced, if commercially available (NOP §205.204). If a particular variety, quality, form or quantity cannot be found from organic suppliers, serious attempts to find it must be documented. A farmer needs to confirm attempts to source organic seeds with written documentation from the seed suppliers contacted. Documentation could include notes taken when a phone call was made to a seed supplier, or copies of order forms requesting organic seeds. Most certifiers expect at least three suppliers that offer organic seeds to be contacted. Organic resource directories, organic farmer organizations, ATTRA, the Organic Materials Review Institute (OMRI) and certifying agents have information on sources of organic seed. All seed must be untreated and non-GMO.

Buffer protects organic crop from non-organic.

Appropriate weed, pest and disease management strategies also need to be implemented (NOP §205.206). Before any approved products can be used, the farmer must use cultural, mechanical and physical methods to enhance crop health, and prevent weed, pest and disease problems. These methods include (but are not limited to) sanitation measures to remove disease vectors, weed seeds and habitat for pest organisms; selection of suitable plant species and varieties; introduction of pest predators; mulching with biodegradable materials; mowing; livestock grazing; hand weeding; mechanical cultivation; and flame weeding. Once cultural, mechanical, and physical methods have been implemented, biological, botanical or mineral inputs, and substances on the National List can be used for weed, pest or disease control.

The organic farmer must implement management practices and physical barriers to prevent commingling of organic and nonorganic products and to prevent contact of organic production and handling operations and products with prohibited substances (NOP §205.272). Physical buffers[28] are needed wherever fields and pastures adjoin areas where prohibited substances are applied

Do Not Spray signs help protect organic buffers and crops.

---

[28] *The NOP does not designate a specific size for the buffer. Historically, certifying agents used 25 feet as the minimum buffer but trees, shrubs and other height barriers are also considered when determining the buffer.*

(NOP §205.202(c)). The size of the buffer is site specific and should be wide enough to prevent pesticide drift or run-off contamination. In most instances, 25-35 feet is adequate. If no prohibited products are used by a neighbor, a signed statement from the neighbor eliminates the need for buffers. If equipment to be used on the organic land is also used on conventional crops, the equipment must be cleaned prior to use on organic crops. Storage areas must be thoroughly cleaned. Storage areas for crops grown for human consumption should be free of rodent and bird droppings.

Prohibited practices for crops and livestock operations include the use of sewage sludge or "excluded methods." "Excluded methods" is the NOP term for genetically modified organisms (GMO). The organic dairy farmer needs to insure that all products used in the operation are non-GMO. Examples of products that might be genetically engineered are seeds, Rhizobia inoculants, corn gluten herbicides, some Bt insecticides, ingredients in feed supplements (carriers, Vitamin E derived from soybeans), amino acid supplements derived from livestock by-products or soybeans, vaccines, vitamins, and milk replacer from rBST-treated cows.[29]

Carefully scan products for GMOs, which are not allowed in organic production.

Organic Livestock Practices and Approved Materials

Slaughter animals sold for organic meat products must be under continuous organic management from the last third of gestation. Dairy cows must be under organic management for at least one year prior to milk being sold as organic. For dairy production, a farm is allowed a one time, whole herd transition where animals must be managed organically for one year prior to certification. Once the herd has been converted, all animals, including calves and heifers intended as replacement stock, must be managed organically from the last third of gestation of the mother (NOP §205.236(a)).

Animals must be fed 100% organic feed and approved feed supplements (NOP §205.237). Growth hormones, drugs to promote growth, plastic pellets for roughage, and feed formulas containing urea, manure, meat or bone byproducts are prohibited. Animals must have outdoor access (shade, fresh air and direct sunlight). Ruminants must have access to edible pasture.[30] Animals may have temporary confinement due to inclement weather, animal's stage of production (young calf), or other conditions that could jeopardize animal health or risk soil or water quality. Adequate housing or shelter is required, with clean dry bedding, opportunity for exercise, and designed to reduce potential for livestock injury. It should be noted that arsenate or other prohibited materials for treating lumber must not be used for new installations or replacement purposes where the treated lumber comes into contact with organic livestock (NOP §205.206(f)). This does not include fence posts and building materials that are isolated from

---

[29] Prohibited under §205.603(c).
[30] §205.2 defines "pasture" as "land used for livestock grazing that is managed to provide feed value and maintain or improve soil, water, and vegetative resources."

production areas. Manure management must not contribute to contamination of crops, soil or water by plant nutrients, heavy metals or pathogenic organisms (NOP §205.239).

A preventative health care management plan must be implemented (NOP §205.238). This health plan must address selection of species, appropriate feed ration, appropriate housing, pasture, and sanitation conditions to minimize stress and accommodate the natural behaviors of the animal. Physical alterations are allowed as long as they are done to promote an animal's welfare and methods used minimize pain and stress. Most certifying agents do not allow tail docking of dairy cows. Vaccines are allowed, except GMO vaccines must be specifically listed on the National List. No GMO vaccines are currently on the National List. Antibiotics, growth hormones, and other prohibited products must be discontinued for one year prior to certification of dairy cows.

Ivermectin is allowed as an emergency use parasiticide when preventive management is not effective. For dairy animals, milk must be withheld for 90 days following treatment with Ivermectin. Breeder stock must not be lactating (milk fed to organic slaughter animals) or in the last third of gestation (NOP §205.238(b) and §205.603(a)(12). <u>It should be noted that some organic milk processors prohibit the use of Ivermectin for lactating animals.</u> Prohibited medical treatments, such as antibiotics, must be used if needed to save the life of an animal but the treated animal and its products cannot then be sold or represented as organic.

Ivermectin is allowed only as emergency treatment.

Refer to NOP §205.603 for a complete list of synthetic products approved for use by organic livestock operations. The organic farmer must review each product's list of ingredients to insure all ingredients meet NOP rules. Strychnine is currently the only nonsynthetic (natural) substance prohibited for use in organic livestock production (NOP §205.604). Some materials are only allowed in certain circumstances or with specific requirements (called annotations in the NOP). In general, alcohol, aspirin, electrolytes (without antibiotics), glucose, iodine, hydrogen peroxide, and vaccines are allowed. Chlorhexidine can be used by a veterinarian in a surgical procedure but is also allowed as an ingredient in a teat dip only when alternative germicidal agents or physical barriers have lost their effectiveness. Glycerin is allowed as a livestock teat dip although the method of production must be through hydrolysis of fats or oils only.

The NOP allows the use of oxytocin for postparturition therapeutic applications only, not on a routine basis. <u>Again, some organic processors prohibit the use of oxytocin.</u> Iodine, copper sulfate and mineral oil can be used as topical treatments. Lidocaine and procaine can be used as local anesthetics but require a withdrawal period of 7 days for dairy animals. Hydrated lime is allowed for external pest control, but not allowed to cauterize physical alterations or deodorize animal wastes. Natural mined lime is allowed. Chlorine materials (calcium hypochlorite, chlorine dioxide, and sodium hypochlorite) and ethanol can be used as disinfectants. Chlorine materials and iodines can be used as sanitizers in the milkhouse but not substances such as quaternary ammonia that leave residues that can contaminate milk.

Organic certification requires producers to keep records that disclose all activities and transactions (NOP §205.103). Records must be kept for 5 years. Records include field histories, activity logs, seed and input use and labels, compost production, buffer crop harvest-storage-use-sales, equipment cleaning, conventional production, harvest and storage records, and sales records such as receipts, invoices, and weight tickets. For dairy farmers, records may also include breeding, purchase of animals, feed and feed supplement purchase and use, pastures, health care product use and an animal identification system (NOP §205.236(c)). All sales of organic or conventional products must be tracked back to the fields of production. The certifying agent may provide examples of records or free sample forms can be requested from ATTRA. (Sample ATTRA forms may be found in appendix D.)

Conversion to organic dairy management

Transitioning to organic dairy farming can take over three years. Some farms may find that their land management already complies with the organic standards and that they only have the one time whole herd transition left. The dairy farmer should use the transition time to gather information, identify problems, implement soil building crop rotations and preventative health care systems, and develop and implement the Organic Plan. Although each dairy farm is unique, several common issues have been identified when a conventional dairy farmer is converting to organic dairying.

Problem Solving

One of the most important changes for the conventional farmer is a new way of looking at problem solving. Because most synthetic products are no longer options, the organic dairy farmer must use alternative strategies to respond to the same types of problems the farmer experienced as a conventional farmer. The thinking process includes identification of the specific problem, identification of causal factors (environmental, nutritional, genetic, etc.) followed by consideration of possible solutions to prevent the problem before it occurs. Secondly, a plan should be in place to deal with the specific problem if it does occur. This is one of the keys to successful organic farming.

For instance, if the dairy operation is having a problem with pneumonia in calves (identification of the problem), the conventional farmer's response may be the use of antibiotics. Since antibiotics are prohibited for organic farmers to use except to save the life of the animal (even then, the animal cannot then be certified), the organic dairy farmer can look at housing (types of bedding, ventilation, moisture levels), calving dates in relationship to weather patterns, cow health prior to calving, building up cow immune system prior to calving, building up calf immune system after birth, genetics, nutrition, and other relevant factors that can be manipulated in order to prevent pneumonia in calves. Additionally, the farmer should develop a response to address pneumonia if it does occur. Solutions might include high quality feed, nutritional supplements, herbal medications, essential oils, homeopathy, probiotics, and acupuncture.

Chapter 12 Understanding Organic Certification

Question and Answer

To clarify specific points in the regulation, the NOP has initiated a Question and Answer section on the NOP website, www.ams.usda.gov/nop/Q&A.html. These questions have been excerpted from the NOP's Questions and Answers.

Q: Can I remove organic calves from my farm, raise them conventionally for a year, bring them back to the farm, and then manage them organically for one year prior to the production of organic milk and milk products?

A: No. Livestock or edible livestock products that are removed from an organic operation and subsequently managed on a nonorganic operation may not be sold, labeled, or represented as organically produced. (Section 205.236(b)(1))

Q: Can I sell an organic dairy animal as slaughter stock?

A: Dairy animals that have been under continuous organic management since the last third of gestation may be sold, labeled, or represented as organic slaughter stock. Conversely, dairy animals that have not been under continuous organic management since the last third of gestation may not be sold, labeled, or represented as organic slaughter stock. (Section 205.236(b)(2))

Q: Do all dairy animals have to be organic from the last third of gestation to produce organic milk or milk products?

A: No. Milk or milk products must be from animals that have been under continuous organic management beginning no later than 1 year prior to the production of the milk or milk products that are to be sold, labeled, or represented as organic. Except that, in the case of the conversion of an entire distinct herd, the animals must receive a minimum of 80 percent feed that is either organic or raised from land included in the organic system plan and managed in compliance with organic crop requirements during the first 9 months of the 1-year conversion period. (Section 205.236(a)(2))

Sourcing Organic Feed
Calculate the amounts of organic feed you have versus the organic feed you will need, based on the number of animals in the herd. Do not add feed supplements when determining feed needs. Mineral and vitamin feed supplements are approved inputs and are not calculated as part of the feed percentage. Good records of feed sources, % of each type of feed in the ration, dates started feeding, date started on pasture, and purchased feed invoices and organic certificates for purchased feed are invaluable when getting the herd certified for the first time.

Some farms have some land, particularly pasture and hay ground, that is certifiable and other land that needs one or two more years to transition. These producers may choose to get certified knowing that they will have to sell crops that are not certifiable and purchase certified organic feed to make up the organic feed ration. Factors to consider are feed quality, regional availability of certified organic forages or grains, and increased prices. <u>If you purchase certified organic feeds, verify that the feed is organic by obtaining a copy of the Organic Certificate or a transaction certificate for that specific load.</u> Producers in this instance need to keep good harvest, storage and sales records for crops sold, and records for purchased feed. Determine the dates your fields and crops, including pastures, are certifiable. Determine if the additional expenses of purchased organic feeds are offset by the higher milk price. Be sure to notify the organic milk processor of your potential dates of certification, so you can start selling organic milk as soon as you are certified. This may mean having a contract or a letter of intent from the processor up to 2 years in advance of your certification. Most often the producer can count on a letter of intent at least 6 months prior to certification.

Approved Materials
The certifying agent will review every input used or intended for use on the organic dairy farm. It is a common problem for dairy farmers to be using health care products, feed supplements, or milkhouse sanitation products that contain synthetic ingredients not allowed. Review all products you use or plan to use. Discontinue using any prohibited products. If you are unsure whether a product is approved, ask your certifying agent for clarification. You may need to provide specific ingredient information. To avoid confusion during the inspection, the dairy farmer should get rid of all products that are not currently used or planned for use under organic management.

Salt is a good example. Some bagged salt contains anti-caking agents to help salt flow easily. If the salt contains yellow prussiate of soda or a "flowing agent", this specific salt product is prohibited. Be sure the salt you feed is pure salt or mineral salt, but not salt with flowing agents.

Another example is mineral oil. The NOP allows mineral oil only for topical use or as a lubricant. Mineral oil is often used as a dust suppressant in feed supplement premixes. Review feed supplement labels to be sure all ingredients are allowed.

Certification resources

Farmers interested in learning more about certification and organic management practices can talk to area organic farmers, attend organic farming conferences and field days, and contact certifying agents directly. You can also subscribe to organic publications such as Organic Broadcaster, The Natural Farmer, In Good Tilth, and the NODPA News. To find out more about organic certification, go to the NOP website (http://www.ams.usda.gov/nop/), ATTRA website (www.attra.ncat.org), Rodale's The New Farm website (www.newfarm.org), and MOSES website (www.mosesorganic.org). Northeast Organic Dairy Producers Alliance (NODPA) websites, www.nodpa.com or www.organicmilk.org are also good sources of information.

Recordkeeping for Organic Dairy Production
Harriet Behar

Some of the best management tools on any farm are the records the farmer keeps on their own operation. While generic statistics can be useful, they are no substitutes for the specific on-farm records to track the unique genetic traits of the herd as well as experiments, successes, and failures of techniques or products. A certain problem may not occur more than once every 2-3 years. Having a written record, rather than relying on memory, detailing the possible solution is valuable both in saving time and in decreasing farmer and animal stress. Making decisions based on your own historical reference builds confidence. The written record provides a continuously improving library of self-developed farm resources. Experimentation with techniques, home remedies, or purchased products can be improved based on your own methods of use and implementation. This is especially important in homeopathy and herbal activities, where the decision on which remedy to use in not only based on the health problem, but also the overall "constitution" of the animal. A variety of remedies may be used for the same problem, depending on the "personality", age, body type, breed, etc.

For the certified organic dairy farmer, up-to-date records are integral to maintaining yearly organic certification. Writing down each activity when it happens is the best way to provide yourself with complete records and lessen stress the night before the inspection when you try to remember what you did with each cow throughout an entire year. Records are used to verify organic management on the farm. They can also be used to evaluate the implementation of your organic system plan. Organic certification is a quality assurance system, with documentation of activities and inputs part of that system. Tracking of organic animals, their feed, and all products used is part of the organic guarantee to consumers, which is one reason they pay more for organic products. Part of the organic premium you receive can be viewed as paying for this "extra" recordkeeping required for organic certification.

Recent "mad cow" and "foot and mouth" outbreaks throughout the world have shown us the importance of clear written records and tracking. Nonorganic agriculture is also moving towards more documentation of livestock production, as part of a system to instill consumer confidence in the safety of their food and to implement effective food recalls when a problem is identified.

Remember, the NOP mandates that all documentation be easily understood and audited by the inspector and the certifying agent (NOP §205.103(b)(2)). Therefore, "outsiders" should be able to identify individual animals in your herd and correlate the animals with their records without your help. The organic inspector will review your records. Not only should there be a system, but it should be complete and up-to-date for each animal.

New dairy conversion record

Farmers who are transitioning their herds to certified organic production must track all feed and health products administered for a full year before they plan to sell organic milk. The dates, amounts and types of feed, as well as their percentage of the finished ration should be clear and verifiable with receipts or harvest data. The date when 100% organic feed is being fed to organic production animals must also be clear and verifiable to aid the inspector and certifying agent in determining the date when you can begin to sell organic milk and receive any organic premium. "Spring" or "sometime in midsummer" are not a detailed enough description for this determination. Exacts dates are needed.

Herd health records

Once a farmer is certified organic, all animals who may produce organic milk or may someday be sold for organic slaughter must have clear identification and complete health records (NOP §205.236(c)). Even if the product or method used for a specific "health event" or problem is allowed under organic regulations, it should be documented to verify compliance to organic standards as well as following the organic plan you submitted to the certifying agent. Stripping out the udder for a few days to manage a mild case of mastitis should be documented. Any use of tinctures, probiotics, aloe vera, etc., must also be written down on the individual's record, with the name of the product and how many times it was administered. Notes on the treatment's effectiveness are helpful in making future decisions, especially when trying out a new product.

Herd health records may take many forms.

Feed and housing need to be clearly documented, since the inspector is only present once per year, but must verify the year-round feeding and living conditions of your animals. For those farmers who rotationally graze, documenting observations of the quality of the pastures throughout the season can help you plan for the best grazing efficiency based on the pasture's capabilities and the time of year. Dates when animals are on pasture and how often they are moved through a rotational grazing system are sufficient to prove compliance with pasture requirements (NOP §205.239(a)(2)).

If you raise your own feed, you will need to track the harvests, yields, storage and periodic use of the organic feeds. If you buy feed, keep track of the receipts. This can help you verify your organic feeding program as well as help you monitor the costs associated with your organic milk production. If any feed handling equipment is used for both organic and nonorganic production, cleanouts before each organic use must be documented. A copy of the feed supplier's Organic Certificate or Transaction Certificate is needed for your files to verify that the feed you purchased is certified organic.

Maintaining a variety of production records both by individual animal and by herd help you ensure a healthy and moneymaking herd of animals. Though not required by the NOP, keeping track of the milk sold per head per day, your somatic cell count (many buyers give premiums or subtract $ based on this count), the rolling herd average of milk production for the year, the calving intervals (how many days open, breeding success or troubles) and how much it actually costs you to produce a hundredweight of milk are all useful in helping you improve your bottom line through use of your own on-farm statistics. Culling animals, selling off young stock, bred heifers or milk cows and deciding on what bull to use can all be better managed with the above information.

Examples of records

The National Organic Program requires basic types of records kept in organic dairy production, but there is no specific style of recordkeeping mandated. The best documentation is complete, easily accessible and will be used consistently throughout the year. Some farmers write items down on a calendar in the barn, milkhouse, or feed room. Then they, or another family member, transfer this to individual animal records periodically throughout the year. This is a good way to have older children be more involved in the management of the farm.

A spiral or three-ring notebook with either a full or half page dedicated to each animal can be used to track breeding, birthing and health events. Individual animal cards on a clipboard or in a recipe box or computer spreadsheets, can also be used. Each production animal must have an individual record detailing these activities, as well as be individually identified with ear tags, numbered neck chains, tattoo numbers or distinct body markings that have been drawn clearly on registration papers or other documents. Individual animal records should include source of the animal if off-farm and its organic status when purchased or the dam/sire and birth date. Individual animal identification number or other identification code, along with the cow's breeding information, should be used to track all of the calves from the milk cows. Observations such as ease or difficulty in birthing, retained placentas, mastitis, etc., should be noted and are useful in your future breeding decisions. Health, weight gain and overall thriftiness of the calf can also be useful information to document. Any products or procedures used should be noted. Remember, only animals that are verified to have been under organic management from the last third of their own gestation before birth to slaughter can be sold as organic meat. The records you keep now can open doors to possible organic meat sales in the future.

A wide variety of livestock recordkeeping forms are available from certifying agents and from ATTRA (see appendix D). These can be used as is, or can be used as templates to help you develop your own forms or system to document your organic dairy production methods. Develop a system that is comfortable for you, and get yourself in the habit of using it. Successful certified organic dairy producers use recordkeeping as an integral tool of their overall farm systems.

Chapter 13

Marketing Organic Dairy Products

Infrastructure for the marketing of milk has been well established in the U.S for many years. Milk cooperatives handle the majority of bulk milk, either bottling for consumer use or processing into butter, cheese, yogurt and other dairy products. Privately owned marketing businesses have begun to develop in recent years as well. Due to segregation issues, organic milk rarely flows through the traditional milk marketing stream. Parallel distribution systems have been set up, but are still developing in many parts of the country. Organic marketing cooperatives, private producer-agent companies, custom processing or on-farm processing and direct distribution are the primary ways organic milk is marketed today.

The Consumer Connection
Jody Padgham

The September 2002 report, "Recent Growth Patterns in the U.S. Organic Foods Market," states that "during the 1990's, organic dairy was the most rapidly growing segment of the organic category, with sales up 500 percent between 1994 and 1999." [31] The report goes on to note that organic dairy products (milk, butter, cheese, eggs, yogurt) comprised 0.9 percent of total U.S. dairy sales in 2000, with sales in this category topping $480 million in 2000. Add to this, 106 million dollars in frozen dairy and ice cream, and we see how significant the organic dairy market has become. Sales of organic dairy products have continued at a steady 27 percent growth rate into the 21st century, leading the growth of the organic food market.

Numerous research studies have been done on why consumers choose organic products. Studies consistently show that consumers prefer organic products because they believe they are better for them and a healthier choice than non-

---
[31] *USDA-Economic Research Service, "Recent Growth Patterns in the U.S. Organic Foods Market" Agricultural Information Bulletin Number 777. September 2002*

organic. Desirable taste and environmental issues are the overall second and third reasons for the purchase of organic products. Successful marketers of organic dairy products will emphasize the nutritional, taste and environmental benefits of organic products through labeling, advertising and shelf signs.

The National Organic Program requires organic certification of handlers and processors of organic milk and dairy products to ensure that organic integrity is maintained from producer to consumer. Retail stores do not need to be certified, but should voluntarily follow procedures to ensure organic integrity.

Organic dairy products have received a significant price premium over conventionally produced products. In the years 1996-1999, a range of 50-72 percent premium was paid for organic milk.[32]

With the organic dairy market significantly smaller than that for conventional dairy products, it is critical that producers closely study market opportunity for their end product before venturing into an organic transition. Several organic milk marketing cooperatives and companies now operate in numerous states, allowing producers to ship bulk milk much as has been traditionally done. Privately owned bottlers and cheese producers will also contract for organic milk production.

On-farm bottling and dairy product production coupled with direct-to-consumer sales is also an option for organic producers in most states. Many states have laws governing on-farm processing and direct marketing. Check with the Department of Agriculture in your state to find out the laws in your area.

Cooperatives and Marketing Associations
Jody Padgham and Chad Pawlak

With organic dairy now a staple for large numbers of American consumers, several organic cooperatives, multi-national companies and dairy marketing organizations have developed to collect, process and distribute organic milk and dairy products.

Find your market first

When considering organic dairy production, it is essential that you contact potential market outlets and find out if they can pick up your production once you become certified organic. Organic milk must be segregated from conventionally produced milk at all stages of production, transportation and

---

[32] Lewrene K. Glaser and Gary D. Thompson. "Demand for Organic and Conventional Beverage Milk." Selected Paper presented at the Western Agricultural Economics Association Annual Meetings, June 29-July 1, 2000, Vancouver, British Columbia, Canada.

processing. Not all areas of the country are served by an organic processing or marketing company, although more geographical area is covered each year due to organic market expansion. It is recommended that you contact all of your potential market sources as soon as you decide to go organic. It can take some time to understand pay programs and the benefits of different shipping choices. There may also be a waiting period to get on a bulk organic milk pick-up. Many farmers new to organic find that it is worthwhile to adjust transition time, including full organic feed conversion to cows, to coincide with an estimated date of first pick-up from their chosen co-op or marketing agency.

A few short years ago the organic dairy producer had limited options in where their milk could be shipped if they did not want to build an on-farm processing plant and take on the responsibility of becoming a wholesaler, distributor and marketer of their own production. Today the options for the producer are growing as the market has grown, and it is important that producers who have started their transition understand these different companies, what their mission statements are and how they conduct business. Once the producers understand the differences, they can compare what they feel is important and ship their milk to the company with whom they have the most in common.

As with other milk co-ops or marketing agencies, most organic marketers will sign a contract with producers stating pay price and pay incentives for quality components. Take the time to understand these programs as they may be different from those you have seen in the past. For example, many organic milk marketing companies test for cold growing bacteria to determine shelf life of milk (PI). Producers who have not seen PI quality programs need to understand what this is and how to manage it on their farm.

Organic milk marketing companies

Organic Valley/CROPP Cooperative, which produces Organic Valley branded dairy products along with a number of other private labels, and Horizon Organic Dairy, now owned by DEAN Foods, are the two largest organic milk marketers in the U.S. Both collect raw organic milk throughout most milk-producing regions of the country and distribute processed product nationally.

There are several regional cooperatives and marketing organizations which collect milk and either sell to one of the large groups, to a local processor or do processing and end product marketing. Examples of regional organic milk marketers include Organic Choice LLC, serving WI and MN, Natural by Nature serving Amish dairies in PA, Farmers All Natural Creamery serving Amish and Mennonite in Kalona, IA, Stonyfield Farm in NH, Springfield Creamery in Eugene, OR, Crystal Cream and Butter Co. in Sacramento,

Organic Valley dairy products can be found in many grocery stores.

CA and Wisconsin Organics in WI. Numerous smaller farmer cooperatives have also been established to promote even narrower niches, such as grass-fed butter produced by Minnesota producers using the Pastureland label and the Wisconsin Dairy Graziers Cooperative in Eastern Wisconsin marketing organic grass-fed cheeses. Some of these smaller processors/marketers only take a portion of the milk produced by their customers/members as they work to develop product, distribution and market share.

Some milk marketing companies and coops do not allow you to split your milk production between more than one marketer. If you plan to do some on-farm processing or to begin a small neighborhood cooperative you should talk to those companies who can take the remainder of your milk and find out if they will allow you to ship only part of your production. Some companies encourage this practice; others may ignore it and others have strict policies forbidding it. The excitement of selling your own milk off your farm as a licensed finished good is a dream of many producers, but if you only sell a small percentage of your annual supply, it is very important to know who will pick up the rest of your production.

Pay programs

Pay programs from different organic milk marketing associations tend to be competitive. Producers who are in transition should familiarize themselves with each of the different pay programs. The best way to understand how the pay prices work is to ask each of the companies to meet with you and explain the program, the terms and quality program. A field representative is the best source of this information.

As with conventional milk, marketing base prices do not always equal the mailbox or net price you will see on your organic milk check. Deductions are a part of marketing milk and you need to understand what deductions are involved and how much will they impact your base price. Some of the deductions you should ask about include:

Hauling or pick-up
Seasonality (Spring flush could result in a deduction and Fall could result in an adjustment as supply becomes tighter)
Marketing/Management
Inventory
Equity
Advertising
Organic service fees (fees paid to the company that has certified your dairy)
Applicable state and federal fees
Regional cost of production (this could be either a deduction or premium depending on market conditions and geographic location)
Any farm assignments you request

Some organic milk marketers have a quota system as part of their contract. As a producer you may be required to ship an agreed upon amount of milk each month. Months that you ship more than your quota may result in lost income opportunity as that milk may be sold conventionally. Months when you have less milk than your agreed quota could result in a deduction to your milk check.

Talking with your neighbors is a good way to understand pay programs, but remember that pay programs vary by year, by geographic area and possibly other factors that can confuse even the most astute economist. It is recommended that you check with multiple sources such as your organic certifying agencies, neighbors and the field agents of the respective marketing agencies to understand pay programs.

On Farm Dairy Processing
Jack Lazor, Butterworks Farm

Check your state laws before committing to any kind of on-farm processing. Each state is different as to the basic legal requirements.

My wife Anne and I began processing milk on the kitchen stove in the late 1970's. As homesteaders with four milk cows, we made yogurt, cottage and farmers' cheese, butter and bottled milk in glass jars. Products were delivered door to door to 25 neighboring families. A milk handler's license came in 1984 and ever since then we've been in the yogurt business. We've grown slowly over the years to a milking herd of 45 Jersey cows. Butterworks Farm yogurt and cream is delivered directly to 40 outlets in Vermont and New Hampshire, and to natural foods stores up and down the east coast by a large distributor. We have managed to survive two decades amidst numerous surprises and changing trends. We would like to share a few of the constants and "prerequisites" needed for successful on-farm milk processing.

Milk processing is serious, time-consuming business and can be difficult to coordinate with other normal dairy farm duties. Cheese or yogurt making must happen simultaneously with milking and crop chores. Dairy plant work can be hot, steamy and confining. Hanging over a cheese vat in a warm humid room on a beautiful sunny hay day takes dedication and fortitude. It is important to find out if this is the work one really wants to do. Hopefully, one or two members of a family would want to do the plant work.

What sorts of products to manufacture on the farm is the next consideration. Raw milk cheesemaking requires the least investment because pasteurization equipment is not needed. A hot water boiler, cheese vat and aging room are the bare essentials. Fluid milk requires having a bottle washer, which takes up lots of floor space and gobbles up large amounts of hot water and caustic soda. Cultured products like yogurt need pasteurization, incubation and specialized cup filling machinery. Returns are highest for yogurt, but start-up costs can be prohibitive with even small cup fillers selling for $150,000 or more. Yogurt, cottage cheese and fluid milk production are all governed by the Federal Interstate Milk Shippers Program, which is a part of the U.S Department of Public Health. Visits to other dairies and frequent consultations with regulatory officials are highly recommended.

Farmstead milk products must be of supreme quality. Practice and learn your trade well. There are plenty of ordinary tasting, mass produced dairy products on the shelf, and the consumer must have a reason to choose yours- especially if the price is a bit higher.

Marketing and distribution are the final pieces to the puzzle of successful on-farm processing. To acquire market share, you must convince someone to cease buying their regular brand and try yours. In-store demos help tremendously. Food buyers love to meet the farmer behind their food. If the shopper likes both you and your product, you've earned their loyalty. Keep the customer supplied with exceptional tasting "signature" dairy products at fair prices and success will follow.

Business size and product distribution are the final considerations in this exploration of farmstead dairy possibilities. The organic marketplace has become huge and rather mainstream. Nationally branded organic dairy products are a reality and you must draw your customer base from consumers who are choosing these brands. A small to medium size herd of cows will produce a lot of cheese or yogurt, but probably not enough to supply a major natural foods distributor or a supermarket chain. Consider targeting a select number of retail outlets within a fifty-mile radius of your farm. Food co-ops and mid-sized family run grocery stores will often permit a new producer to do demos on busy shopping days. Forge personal relationships with your retailers and consumers. Good will and taste go a long way. Business will grow slowly and steadily. New outlets can be added as needed. Once you're somewhat established, stores will begin calling you. Organic food and farming are here to stay. Properly managed small-scale organic on-farm dairies have every reason to thrive in local and regional markets.

Appendix A

Organic Matter Research
Walter Goldstein

Managing root health

Results from trials in Wisconsin and Holland suggest that conventionally-grown cereals may invest more of their resources into root production than crops grown with sustainable management systems, and that they may need more roots to form optimal yields.

We studied root health and growth of corn on 17, 12, and 16 farms in 2000, 2001, and 2002 in Wisconsin, Iowa, and Illinois. About half of the farms were conventional and half were organic. Averaged over the three years, the conventional corn had root disease scores of 26% while the organic systems had scores of 15%. The highest disease incidence was found where corn followed corn (30%) and the least disease was found where corn followed organic soybeans (15%). It is possible that the reason for these differences has to do with the use of organic manures, which can induce suppressive conditions for disease organisms.

Disease stimulates production of roots

In trials conducted in 2001 and 2002, corn grown after corn or soybeans in conventional systems produced more roots (5232 lbs/acre) than corn grown in the organic systems after soybeans or forages (4,442 lbs/acre), possibly to compensate for poorer soil quality and greater root disease problems. On average, corn grown conventionally after corn and soybeans on 27 sites had root/bushel ratios of 65:1, while corn grown organically after forages or soybeans on 53 sites had root/bushel ratios of 38:1.

Health causes more efficient use of nitrogen

By shifting its resources away from grain production and towards root production, this corn apparently also mineralized more nitrogen from soil organic matter, took up significantly more nitrogen, and needed significantly more nitrogen for every bushel of grain produced than did the other systems. Corn grown after corn or after soybeans in conventional systems took up 1.8 lbs of nitrogen for every bushel of corn produced. Corn grown after organic forages or soybeans took up only 1.4 lbs of nitrogen per bushel produced. The conventional corn appears to be less efficient.

Green manure systems

Corn grown after small grains with under-seeded green manure legumes such as red clover had low root and grain production and also showed poor nitrogen efficiency. This green manure system seemed to be associated with a lowered ability for corn to compete with weeds and with a lowering of yield potential. The most efficient systems for transforming soil organic nitrogen into grain were where corn followed after alfalfa, alfalfa + grass or after soybeans in an organic rotation that included perennial forages and routine applications of animal manure.

Long term effects of compost and biodynamic growth regulators

On the other hand, some of our on-farm and plot research suggest that the long-term use of dairy manure and compost may stimulate root production. Furthermore, natural growth regulators used by biodynamic farmers seem to also increase root formation. Long-term trials in Wisconsin at Michael Fields Agricultural Institute suggested that the enhanced root production associated with biodynamic growth regulators may lead to more stable yields.

Corn's contribution to organic matter and soil quality

Corn produces large quantities of roots and stalks and may thereby make a considerable contribution to organic matter of soils. In general, the quantity of corn stalks is equivalent to the grain yield, but the amount of roots produced varies greatly from farm to farm. However, the organic matter that comes from corn roots may not have a strong beneficial effect on soil structure. The long-term effect of corn roots on soil appears to be blocky soil structure, with a predominance of welded rather than biogenic aggregates. Earthworms are relatively discouraged by rotations with lots of corn. Some evidence even exists that dead corn roots may inhibit microbial activity in the soil. Though the residues from soybeans may create a more friable soil structure, they do not have a substantially greater positive effect on the stability of soil crumbs.

On-farm trials in 2000 suggested that corn produced 5,000 to 13,000 lbs/acre of roots, with the actual amounts relating roughly to yield. This may correspond to 2,000-5,200 lbs of organic-carbon. The crucial questions are: how much of this becomes soil organic matter and what do these corn residues do for the soil? Studies in the Midwest suggest that 2-11% of the carbon in corn stalks may become organic matter, while corn roots may contribute up to 18%. Canadian estimates suggest that about 13-22% of the C in corn residues may be retained for medium textured soils, less for sandy soils and more for clays.

The contribution of perennials to organic matter

Studies of native prairies have shown that they invest more of their resources into producing roots and soil organic matter than our other crops. However, it is difficult to use these prairie grasses in modern farming. Mixtures of alfalfa and perennial grasses also have beneficial effects on soil organic matter, soil structure and the accumulation of young active organic matter. These benefits may increase as the stand ages, because there are more roots and more lignin in the roots. Mixing alfalfa and grass plants together is a good idea for many reasons. The alfalfa fixes large quantities of nitrogen, and some of this ends up fertilizing the grasses. In fact, alfalfa plants have to fix more nitrogen for themselves when they grow with grasses than if they grow alone. Furthermore, alfalfa and grass plants feed organic matter and nitrogen to the soil while they are growing. This is because they produce large quantities of decomposing roots, and because large quantities of leaves are lost to the ground while making hay.

Root turnover

Research suggests that about half of the total roots that are formed by an alfalfa plant may die and turnover during a growing season. Furthermore, as the alfalfa stand thins out the dying roots feed the soil. Grasses can accumulate much larger quantities of roots than alfalfa, while still contributing a minor portion of the hay yield. They also have strong beneficial effects on forming stable, biogenic crumbs. If the quantities of decomposing roots and leaves are figured into the equation, the incorporation of high yielding, established stands of alfalfa and grass may add 5,000 to 7,500 lbs of organic carbon to the soil. Roots of alfalfa decompose quickly in the soil, releasing large quantities of nitrogen, but roots of perennial grasses decompose slowly and help maintain soil structure.

Grazing

Under management-intensive grazing conditions grass and legume plants may not have sufficient reserves to replenish roots, which tend to die following a grazing event. Heavily grazed pastures tend, over time, to form a dense mat of roots in the top three to four inches of soil with little root development below this zone. Such root development can lead to poor production in drought years. There appears to be a general relationship between how high the grass plant is allowed to grow and its rooting depth. Ways to avoid this decline in pasture productivity are to rest the pasture or to break up the pasture and annual crop for a year or two.

Appendix B

Vitamin Function and Deficiency in Livestock

Vitamin	Function	Results of Deficiency
A (retinol)	Vision cycle-adaptation to light/dark	Night blindness
Provitamin A (carotene)	Tissue growth and repair, especially skin and mucous membranes, normal growth and development of bones, reproductive cycle, lactation, mobilization of iron stores in liver	Susceptibility to tissue infections poorly function tissue, rough, dry skin, dry mucous membranes cells, nerve damage, failure to grow, poor, reproduction, lactation, anemia
The B-complex vitamins		
Thiamine (B1)	Normal growth, carbohydrate metabolism, normal function of heart, nerves and muscles	Loss of appetite, gastric distress, deficient digestive juices
Riboflavin (B2)	Normal growth and vigor, protein and energy metabolism	Poor wound healing, eye irritation and sensitivity to light, skin eruptions
Niacin	Energy production, normal growth, health of skin, normal activity of stomach, intestines and nervous system	Weakness, lack of energy, poor feed consumption and conversion
Pyridoxine (B6)	Amino acid metabolism: protein synthesis, heme formation, brain activity	Anemia, hyperirritability, poor tissue repair, hormone imbalances
Pantothenic acid	Fat, cholesterol and heme formation and amino acid activation	Lack of energy, anemia, poor tissue repair
Folic acid	Growth and develop red blood cells	Certain types of anemia
Cobalamin (B12)	Normal red blood cell formation, nerve function, and growth	Pernicious anemia, irritability, poor growth
D (calciferol)	Absorption of calcium and phosphorus calcification of bones	Faulty bone growth, poor bone development
E (tocopherol)	Antioxidant, normal growth, reproduction, interacts in many enzyme systems, general protective role in cellular and subcellular membrane structures	Poor utilization of Vit. A, abnormal breakdown of red blood cells, poor growth patterns, reproduction problems
K (menadione)	Normal blood clotting	Bleeding tendency, hemorrhages

Mineral Function and Deficiency in Livestock

Mineral	Function	Results of Deficiency
Calcium (Ca)	Bone formation, blood clotting, muscle contraction and relaxation, heart action, nerve transmission	Soft bones, slow-clotting blood, irritability, weak or irritable heart muscle
Phosphorus (P)	Bone formation, overall metabolism of carbohydrates and fats	Poor growth, poor rate of gain
Sodium (Na)	Water balance, osmotic pressure, acid-base balance glucose absorption, muscle action	Imbalances in water shift and control, imbalances in buffer system, gastrointestinal disorders
Potassium (K)	Water balance in cells, acid-base balance, muscle and nerve action, protein synthesis	Water imbalance, irregular heart action, tissue breakdown-potassium loss
Iron (Fe)	Hemoglobin formation	Anemia, poor growth patterns
Iodine (I)	Synthesis of thyroid hormone (which regulates metabolic rate), cell oxidation	Impaired metabolic rate with inability to meet demands of high producers
Magnesium (Mg)	Normal heart function, energy production, muscle and nerve maintenance calcium and phosphorus metabolism	Poor feed consumption, irritability of all muscles including heart, hyperexcitability
Sulfur (S)	Blood clotting, bone formation, muscle metabolism, neutralization of toxins	Blood clotting problems, build up of body toxins, generalized weakness
Chlorine (Cl)	Maintenance of fluid balance, formation of hydrochloric acid	Fluid imbalances, inefficient digestion
Copper (Cu)	Formation of red blood cells, enzymes	Anemia, insufficient hydrochloric acid production
Manganese (Mn)	Essential in several enzyme systems, esp. bone development and insulin	Poor bone development and maintenance, diabetes
Cobalt (Co)	Formation of red blood cells	Anemia
Zinc (Zn)	Reproduction, production of many essential enzymes	Inadequate reproductive cycle, poor growth patterns
Molybdenum (Mo)	Iron metabolism	Iron deficiencies/anemia
Selenium (Se)	Reproduction, metabolism	Retained placentas, poor fertility/growth

Appendix C:

Alternative Dairy Breeds
(Other Than Holsteins, Jerseys, and Swiss)
By Robert Hadad, HSUS

Information provided by the American Livestock Breed Conservancy www.albc-usa.org

Dexter

Dexter cattle are one of the smallest breeds of cattle. They stand about 40" tall and weigh from 700 lbs. to 900 lbs. Due to their size, Dexter cattle are ideal for small farms. The breed was developed in the 1800s in southern Ireland from the Kerry dairy breed. It is a dual-purpose animal having improved beef qualities as well as being good for milking.

The Dexter cattle are solid and strong, sometimes even used as a draught animal. Coloration is usually black but red and dun variations can sometimes be seen. The cattle have black-tipped horns that point upward. Two body types are found within the breed. There are cattle with normal bodies and very short legs. This type can sometimes produce non-viable offspring. The second type is proportionally smaller in every dimension.

Dexter cattle are hardy and maternalistic. They are forage efficient and their good browsing characteristics can help rid pastures of invasive weed pests. Dexter's make a good dual-purpose dairy or beef animals. The milk produced depends on whether the lines of animals were selected for beef or milk. The milk from dairy lines is high in milk solids, ideal for butter and cheese production.

Red Poll

Another dual-purpose breed, the Red Poll, comes from England and dates back to the early 1800s. Breeding over the years selected for good beef and dairy qualities resulting in animals that have consistent traits, are dark red, and are always polled. Animals are known for longevity and high milk production.

Red Poll cattle are medium sized animals weighing 1,200 lbs. for cows and up to 1,800 lbs. for bulls. They have quiet dispositions, ideal for small scale dairy operations as well as larger herds. They are easy to handle making Red Polls desirable for rotational grazing systems. The milk quality is high in protein and butterfat, ideal for cheese production.

Dutch Belted

The Dutch Belted breed is named after its country of origin and the color pattern of the animals. The cattle are black on the front and back end with a white belt around their middles. The Dutch Belted, known as the Lakenvelder in the Neterlands, is a very old breed dating back to the 1500s where, even then, it was noted as a fine dairy breed.

After nearly being totally replaced in the United States by Holsteins in the 1970-80s, the Dutch Belted is being sought after once again. Farmers interested in grass-based dairying find the Dutch Belted well suited, and with good quality forage, they will maintain condition, producing a respectable amount of milk from grass grazing. Cows weigh around 1,200 lbs. and bulls can weigh upwards of 1,700 lbs.

provided by Dutch Belt Cattle Association of America

Milking Shorthorn

The Shorthorn is a famous breed and one of the first to be improved. The coloration of red, white, and roan made this breed easily recognizable as it spread across the country. In the 1820s, Shorthorns were concentrated in Ohio and Kentucky where large ranges of grass were available as well as plenty of corn. The Shorthorn was easily adaptable to both dairy and beef production. As the use of Holsteins increased, dairy Shorthorns lost popularity while beef Shorthorns still maintained a strong interest.

The dairy Shorthorns were crossed with various breeds like the Australian Shorthorn and the Red and White Holstein to gain greater milk productivity. What was lost was the breed's ability to do well on pasture. Still, in the U.S., some purebred strains remain. These strains show good performance on a grass-based dairying system. They also forage efficiently, are healthy, long-lived, and productive plus having the added benefits of good beef quality.

provided by Hoard's Dairyman

Guernsey

Guernsey cattle originated on the island of Guernsey, which is found among the Channel Islands between England and France. The breed is an old one dating back before the 1700s. Guernsey cattle were selected by breeders for color consistency (fawn to gold range). Their size is medium with cows weighing 1,400 lbs. and bulls reaching over 2,000 lbs. The Guernseys are horned, but polled strains have been developed. Guernsey cattle can do well on good quality pastures and in a rotational grazing system and will produce a moderate amount of milk. Their milk contains high levels of carotene, which is a precursor of Vitamin A and responsible for giving their milk its distinctive golden color. Golden butter is a value added benefit of dairying with Guernseys as it is preferred by chefs and bakers for fine recipes.

Canadienne

This breed of cattle is one of the truly historic breeds developed in North America. It originated from cattle brought over from Normandy and Brittany to Canada in the early half of the 1600s. The breed was selected for its adaptation to the rigorous environment of Eastern Canada. The cattle have distinctive coloration of black, brown, tan, or russet with a fawn colored muzzle and udder. They have black-tipped horns. The breed is not very large, with cows weighing about 1,100 lbs. and bulls weighing about 1,600 lbs.

Canadienne cows are productive even on poor forage or challenging conditions. Milk production can stretch over 300 days and the milk can have a butterfat content of 4.3% and a protein level of more than 3.5%. Over the years, purebred herds have been diluted and the number of animals has been reduced from once high levels. The Canadienne are ideal animals for grass-based grazing systems with their lengthy productivity and adaptability to rough conditions.

Ayrshire

This breed was purposely developed by Scottish farmers to increase milk and meat production. Further breeding increased the quality of milk for cheese and butter production. The breed was officially recognized in the early 1800s.

Ayrshires are alert and have active dispositions. Cattle are white with dark red or reddish brown spots and speckles. Some have brown legs. Cows can weigh up to 1,100 lbs. and bulls can weigh 1,600 lbs. or more.

Besides being good-looking animals, Ayrshires are very hardy animals well adapted to harsher environments with colder climates. Because of this, the cattle are able to do well on forage and make excellent grazers.

provided by Hoard's Dairyman

Chapter 14 Appendices

Appendix D:

Livestock Record Keeping Forms

These sample forms have been provided by ATTRA- Sustainable Agriculture Resource Service. They may be copied and used as needed. You may get additional forms from ATTRA by calling 800-346-9140 or on the web at www.attra.org/attra-pub/livestockforms.html

ATTRA offers

D.1. Pasture/Paddock/Hayfield Activity Log

D.2. Individual Organic Animal Health Record: Large Livestock

D.3. New Organic Dairy Herd Feed Record

D.4. Breeding Record: Large Livestock

D.5. Feed Supplements and Additives: Product Inventory

D.6. Organic Livestock Sales Summary

D.1. Pasture/Paddock/Hayfield Activity Log

Organic Farm Documentation Forms

Provided courtesy of NCAT's ATTRA Project, 1-800-346-9140

Pasture / Paddock / Hayfield Activity Log

A record of the practices and equipment you use for each pasture or paddock and hayfield.

Farm Name or Unit: _____ Field I.D.: _____ Acres: _____ Year: _____

Forage(s):

Date(s)	Activity	Date(s)	Activity

Forage / Hay Harvest: (e.g. tonnage, bales, animal units, etc.)

Date(s)	Yield	Date(s)	Condition of Forage or Hay

Additional notes and observations:

original Field Activity and Inputs Log p.1

original FIELD ACTIVITY and INPUTS LOG				
A record of the practices, equipment, and materials you use for each individual field.				
Producer Name:		Field Number:	Crop Year:	

FIELD PREPARATION: List date and activity, e.g. chisel plowing, moldboard plowing, discing, etc:

Date	Activity	Date	Activity

PRE-PLANT fertilization/soil amendments, pest control treatments, etc:

Date	Activity	Material Applied	Rate/Amount	

Notes and observations:

PLANTING:

Date	Field/Bed Number	Crop/Variety Planted	Seeding Rate	Condition of Stand	

PLANTING and POST-PLANTING fertilization/soil amendments, pest control treatments, seed inoculants/treatments, etc:

Date	Activity	Material Applied	Rate/Amount	

Notes and observations:

POST-ESTABLISHMENT FIELD ACTIVITIES: Cultivation, mowing, hoeing/weeding. Describe crop condition as excellent, good, average, or poor.

Date	Activity/Equipment	Crop Condition	

Notes and observations:

PEST MONITORING: List date, specific field number, type of insect or pest, and assessment of crop damage you observed.

Date	Insect/Pest	Type of crop damage	Damage Assessment (Low, Medium, High)

Chapter 14 Appendices

original Field Activity and Inputs Log p.2

DISEASE MONITORING: List date, specific field number, type or description of disease, and assessment of crop damage you observed.

Date	Disease	Type of crop damage	Damage Assessment (Low, Medium, High)

WEED MONITORING: List date, specific field number, name or description of weed, and assessment of crop damage you observed.

Date	Disease	Type of crop damage	Infestation Level (Low, Medium, High)

HARVEST MONITORING: Use harvest/storage records to provide more detailed harvest information.

Date	Yield	Condition of Harvest	

POST-HARVEST FIELD ACTIVITIES: e.g. planting of cover crops, manuring, shredding crop residues, etc:

Date	Activity	Date	Activity

Notes and observations:

Input Details:
Record input data in this section. Indicate whether input is approved for organic use (A), use is restricted (R), or use is prohibited in organic agriculture (P)

Purchase Date	Input Material	Brand/Source	Organic Status:(A) (R) (P)	Date of Application	Rate of Application

Chapter 14 Appendices

D.2. Individual Organic Animal Health Record: Large Livestock

Organic Farm Documentation Series

Provided courtesy of NCAT's ATTRA Project, 1-800-346-9140

Individual Organic Animal Health Record: Large Livestock

Preventive health care practices must be used, including appropriate species selection, complete feed ration, good housing, outdoor access, exercise, needed physical alterations, and vaccines. Synthetic medications, including parasiticides, must be on the National List. You must not withhold medical treatment to preserve organic status; all livestock treated with prohibited substances must be clearly identified. [§205.103, §205.238, §205.603]

Farm Name or Unit: _____ Production Year: _____

Tag#, ID#, or Name of Animal _____ *and/or* Description of Animal _____

Birth Date _____ Dam I.D. _____ Sire I.D. _____ Sale Date _____ Buyer Name _____

If the animal died, what was the cause of death? _____ Date of Death _____

Vaccinations / Date Administered	Vaccinations / Date Administered

Physical Alteration / Date	Physical Alteration / Date

Health care problem	Practice(s) used to treat health care problem	Product(s) used to treat health care problem	Date used / Effectiveness

D.3. New Organic Feed Record

Organic Farm Documentation Series

Provided courtesy of NCAT's ATTRA Project, 1-800-346-9140

New Organic Dairy Herd Feed Record

Use this form to calculate percentages of organic and non-organic feed rations for new (converting*) dairy herds only. Include ration information on feed supplements and additives, such as salt or mineral mixes. Indicate dates ration was fed, as feed rations change seasonally. Use a new form for each class of animal. [§205.236(a)(2)]

Farm Name or Unit: _____

Production Year: _____

Check type of dairy animal receiving this ration: Production females _____ Replacement females _____

Feed (pasture**, hay, silage, corn, grains, processed feeds, etc.)	% of daily ration	Organic (O) Transitional (T) Conventional (C)	Source of feed (field ID for farm-grown)	Date you began feeding this ration	Date you discontinued this ration (if applicable)

*When an entire, distinct herd is converted to organic production, the herd must be fed a minimum of 80% certified or certifiable organic feed for 9 months, followed by 100% organic feed for the final 3 months of the conversion period, and 100% organic feed thereafter. See §205.236(a)(2) of the National Organic Regulations.

**Pasture history information can be documented in the Pasture Access Form.

D.4. Breeding Record: Large Livestock

Organic Farm Documentation Series

Provided courtesy of NCAT's ATTRA Project, 1-800-346-9140

Breeding Record: Large Livestock

Keep breeding records for all animals used in the production of organic meat and dairy products, and for stock that may be sold as organic breeding stock or organic slaughter stock. Sire information is not required. [§205.236]

Farm Name or Unit: _____   Production Year: _____

Bred Female ID	Sire ID	Freshening Date	Offspring ID	Parasiticide Use? Provide product name and date of last use, if appropriate	Disposition of offspring (e.g. kept as replacement stock, sold as organic, sold as non-organic)

*All synthetic parasiticides are prohibited in slaughter stock. See the National List for other uses. [§205.238(b), §205.603(a)(12)]

D.5. Feed Supplements and Additives: Product Inventory

Feed Supplements and Additives: Product Inventory

Track your current inventory of feed supplements and additives with this form. The producer keeps this ongoing list to document specific products currently in use for organic livestock and the date they were approved by the certifying agent, if needed. Keep copies of labels on file. [§205.237]

Farm Name or Unit:

Production Year:

Feed supplement/ additive	Source or brand	Date of first use	Organic?	Item on National List*?	Approved by certification agent? (include date approved)	Reason for use	Date discontinued use

* The National List of synthetic substances allowed for use as feed supplements and feed additives in organic livestock production can be found under §205.603(c) and §205.603(d).

Organic Farm Documentation Series   Provided courtesy of NCAT's ATTRA Project, 1-800-346-9140

… Chapter 14 Appendices

D.6. Organic Livestock Sales Summary

Organic Farm Documentation Series · Provided courtesy of NCAT's ATTRA Project, 1-800-346-9140

Organic Livestock Sales Summary
[§205.103(b), §205.236(c)]

Farm Name or Unit: _____ Production Year: _____

Type of Livestock Sold: _____

Date of Sale	Animal I.D.	Organic / Transitioned / Conventional	Sale Price	Bill of Sale Number	Buyer	Reason for Sale

Glossary

- **Acidosis:** An abnormal increase in the acidity of the body's fluids, caused either by an accumulation of acids or the depletion of bicarbonates. In a ruminant generally caused by feeding too much grain.
- **ADF:** Acid Detergent Fiber. A measurement of fiber that can be extracted with an acidic detergent. Includes cellulose, lignin, ADIN and acid-insoluble ash. ADF is highly correlated with cell wall digestibility. The higher the ADF, the lower the digestibility or available energy.
- **Antibacterial:** Destroying or inhibiting the growth of bacteria.
- **Approved Materials:** Allowed for use in organic production and handling under the National Organic Program.
- **Calf scours:** Diarrhea of young animals. May have several causes.
- **Casein:** A component of milk protein, critical in the production of cheese.
- **Certifying Agent:** Independent third party organization accredited by the USDA to approve organic certification under the National Organic Program.
- **Colostrum:** The thin fluid secreted by the mammary glands at the time of birth. Rich in antibodies and minerals, precedes the production of true milk.
- **Conventional farm/system:** Non-organic farm/system
- **Crop rotation:** The practice of alternating different types of crops in the same field or area. Usually from one growing season to the next, could be several rotations within one year for some systems.
- **Cross Breeding:** Breeding two animals of different pure breeds.
- **Dairy Conversion:** The act of switching a dairy farm from conventional management to organic management.
- **Displaced abomasum:** When the abomasums (true stomach) migrates from the bottom right side of the cow up, flips and fills with an air cap. "Twisted stomach"
- **Dry Matter:** The part of a feed which is not water. Calculated by figuring the percentage of water and subtracting the water content from 100%.
- **Euthanasia:** The act of causing a painless death.
- **Forage:** Vegetative material in a fresh or ensiled state (pasture, hay or silage) which is fed to livestock. *(From the NOP)*
- **Foundation breeding:** A system of breeding whereby individual farms develop on-farm pure breed breeding systems to optimize qualities specific to those farms. Purebred lines may then be utilized by other farmers seeking new bloodlines.
- **Freshening:** To calve and then begin to produce milk.
- **Functional Traits:** Genetic traits such as longevity, disease resistance, ability to become pregnant. Useful in the life of the animal.
- **GMO: (Genetically Modified Organism):** Organisms that have been changed through one of several kinds of gene manipulation, including cell fusion, microencapsulation and macroencapsulation and recombinant DNA technology.
- **Grazier:** Someone who uses a MIG system to manage their animals and pasture.

Chapter 15 Glossary

- **Inputs:** Products used on the farm that must be brought in from off-farm. Includes medicines, nutrients, control products.
- **Ivermectin:** Common parasiticide. Allowed in organic systems only when other treatments are proven ineffective.
- **Mastitis**: Inflammation of the udder.
- **MIG:** Managed Intensive Grazing. The process of frequently moving animals into a series of managed paddocks for grazing. The goal of MIG is to increase level of nutrients available to animals through intensive pasture management.
- **National Organic Program:** The Program authorized by the Organic Foods Production Act of 1990 for the purpose of implementing its provisions. *(NOP)*
- **NDF:** Neutral Detergent Fiber. A measurement of fiber after digesting in a nonacidic, nonalkaline detergent. Used to determine forage quality. Contains the same fibers in an ADF reading, plus hemicellulose.
- **NEL:** Net Energy for Lactation. The estimated energy available for milk production, as measured in calories.
- **Nosode:** (from nosos, the Greek word meaning disease) a homeopathic preparation made from matter from a sick animal or person. Substances such as respiratory discharges or diseased tissues are used. The preparation, using alcohol, repeated dilution and succussion, essentially renders the substances harmless, while producing a powerful remedy. The use of nosodes for preventing disease has been employed in veterinary and human homeopathy for many years.
- **Organic Certification:** The process of third party accreditation of a livestock, crop or handling operation in which conformation to the organic rule as described by the National Organic Program is assured.
- **Organic matter:** The remains, residues or waste products of any organism. *(from the NOP)*
- **Resistance traits:** Inherent ability of an animal to withstand specific pressures, such as from disease, parasites.
- **Rumen:** An extremely large compartment in a ruminant's digestive system, acts as a large mixing and fermentation vat.
- **Preventative Health Care:** The act of improving the base health of an animal so as to avoid critical health issues. Also anticipating health challenges and pre-treating before symptoms occur (i.e. giving immune boosting aides before stressful situations).
- **Subcutaneous:** Under the skin. With a subcutaneous injection, a needle is inserted just under the skin. A drug can then be delivered into the subcutaneous tissues. After the injection, the drug moves into small blood vessels and the bloodstream. The subcutaneous route is used with many protein and polypeptide drugs such as insulin which, if given by mouth, would be broken down and digested in the intestinal tract.
- **TDN:** Total Digestible Nutrients. The estimated difference between food going in and manure coming out.
- **TMR:** Total Mixed Ration
- **Transition:** Period of time that organic management practices are being followed but organic status is not yet approved.

Resources

Soil, Crops, Pest Biology and Diversity Practices

ATTRA publications, including the organic field crop production guides and the series on pest management. Available for downloading from www.attra.org or by request (1-800-346-9140)

A Whole-Farm Approach to Managing Pests. Sustainable Agriculture Network, Hills Bldg., Room 35, University of Vermont, Burlington, VT 05405-0082. Available online at http://www.sare.org/publications/farmpest/farmpest.pdf

Bowman, Greg. *Steel in the Field* 2002. Sustainable Agriculture Network, Hills Bldg., Room 35, University of Vermont, Burlington, VT 05405-0082

The Organic Field Crop Handbook, 2000. Canadian Organic Growers, Box 6408, Station J, Ottawa, Ontario, Canada K2A 3Y6 613/231-9047.

Bowman, Shirley and Cramer, *Managing Cover Crops Profitably, 2nd edition.* 2001. Sustainable Agriculture Network. Available online at http://www.sare.org/publications/covercrops/covercrops.pdf

McCaman, Jay. *Weeds and Why They Grow,* 1994. Jay L. McCaman, Box 22, Sand Lake, MI 49343. 616-636-8226

Zimmer, Gary. *The Biological Farmer,* 2000. Acres USA, PO Box 91299, Austin, TX 78709. www.acresusa.com, 1/800-355-5313

Magdoff, Fred and Harold van Es *Building Soils for Better Crops, 2nd edition.* 2000 Sustainable Agriculture Network. Available online at http://www.sare.org/publications/bsbc/bsbc.pdf

Soil and Water Conservation Society (SWCS). 2000. *Soil Biology Primer.* Rev. ed. Ankeny, Iowa: Soil and Water Conservation Society. Available online at http://soils.usda.gov/sqi/soil_quality/soil_biology/soil_biology_primer.html

Certification

Jannasch, Rupert and REAP-Canada staff, Introduction to Certified Organic Farming, Ottawa: Canadian Farm Business Management Council, 2nd edition, 2002. ISBN 1-894248-23-1

An Organic and Sustainable Practices Workbook & Resource Guide for Livestock Systems. National Center for Appropriate Technology (NCAT), April 2003.

Kuepper, George, et al, *Organic Livestock Documentation Forms.* National Center for Appropriate Technology (NCAT), August 2003. Available online at www.attra.ncat.org

The Upper Midwest Organic Resource Directory. Midwest Organic and Sustainable Education Service, 5th Edition, 2004. Available online at www.mosesorganic.org

Richards, Steve, et al. *The Organic Decision: Transitioning to Organic Dairy Production.* Ithaca, New York: Cornell University, 2002.

Veterinary Care

Dettloff, Paul D.V.M., *Alternative Treatments for Ruminant Animals.* 2004 ACRES USA, Austin, TX 2004.

Duvall, Jean. *Control of Internal Parasites in Cattle and Sheep,* Quebec, Canada. Publication #70, Ecological Agriculture Projects, Macdonald College 1997. Available online at http://www.eap.mcgill.ca/general/home_frames.htm

Duvall, Jean. *Treating Mastitis Without Antibiotics* Publication #69, Ecological Agriculture Projects, Macdonald College 1997. Available online at http://www.eap.mcgill.ca/general/home_frames.htm

Hoke, David, *A New Troubleshooter's Guide to Dairy Cows.* NOFA-VT 1994. PO Box 697, Richmond, VT 05477

Karreman, Hubert J. V.M.D. *Treating Dairy Cows Naturally: Thoughts and Strategies* Paradise Publications, Paradise PA 2004.

Macey, Anne. Editor. *Organic Livestock Handbook* Canadian Organic Growers, Ottawa, Ontario, Canada 2000.

Sheaffer, C. Edgar. V.M.D. *Homeopathy for the Herd, A Farmer's Guide to Low-Cost, Non-Toxic Veterinary Care of Cattle, Dairy & Beef.* ACRES USA, Austin, TX 2003.

Yasgur, Jay. *Yasgur's Homeopathic Dictionary and Holistic Health Reference, 4th Ed.* 1998. Van Hoy Publishers PO Box 636, Greenville, PA 16125

Economics and Marketing

Born, Holly. *Enterprise Budgets and Production Costs for Organic Production* ATTRA, National Sustainable Agriculture Information Service, July 2004.

Dairy Your Way - information and farm profiles that help farmers compare and consider livestock production options for their enterprises. MN Dept of Agriculture, early 2005

Dunaway, Vicki. *The Small Dairy Resource Book*, Sustainable Agriculture Network. Out of print, available online at http://www.sare.org/publications/dairyresource/dairyresource.pdf

Greenbook, Minnesota Department of Agriculture. 2001, 2002, 2003, 2004, 2005. www.mda.state.mn.us/esap/Greenbook.html

McCrory, Lisa. *An Economic Comparison of Organic and Conventional Dairy Production, and Estimations on the Cost of Transitioning to Organic Production.* Northeast Organic Farming Association of Vermont. May 2001.

Pasture and Grazing

Gerrish, Jim. *Management-intensive Grazing. The Grassroots of Grass Farming.* Stockman Grass Farmer PO Box 2300, Ridgeland, MS 39158. 1/800-748-9808

Nation, Allan. *Quality Pasture, How to create it, manage it, and profit from it.* Green Park Press, Stockman Grass Farmer PO Box 2300, Ridgeland, MS 39158. 1/800-748-9808

Voisin, Andre. *Grass Productivity.* 1988. Island Press, PO Box 7, Covelo, CA. 95428. First published in 1959 by Philosophical Library, Inc.

General

Benson, Laura and Zirkel, Robert. *Organic Dairy Farming*, 1995. Community Conservation, 50542 One Quiet Lane, Gays Mills, WI 54631. Available online at http://www.communityconservation.org/store.htm#odf

Index

A

acetic acid 13, 14, 67
acidosis 8, 26, 46
acid degree value 71
Acid Detergent Fiber 11–12
acid sanitizer 67
acres, certified organic in the U.S. 3
acupuncture 144
adjoining land use 137
alcohol 25, 44, 143
allelopathic 103, 112, 117
All Natural Creamery 153
aloe 26–27, 44–47, 148
anemia iv–v
antibacterial 43
antibiotics 31, 37, 41, 43, 47, 55, 72, 143–144
  in calf milk 36
  in culture interpretation 48
  resistance to 49
  resistance to natural 43
  testing for 61, 63
arnica 44
artificial insemination 83
aspirin 143
ATTRA 141, 144, 147, 150
  forms xi

B

bacteria 48, 60–63, 69, 109, 125
  and milk quality 59, 153
  as mastitis cause 48–49
  in intestine 36
  in rumen 8–9
  in silage inoculants 13
  in soil 86, 91, 132
  preventative measures 64, 67–69, 70–72
Bangs 62
bedding 29, 33, 57, 62, 64, 71, 74, 84, 89, 142, 144
behavior 52
biodiversity 2, 7, 79, 132–133, 140
biosecurity 5, 76–78, 81
bloat 126
Body Condition Score 16, 19
botanicals
  as pesticides 116, 141
  in health care 27, 44, 46

Bovine Spongiform Encephalopathy 55
breeding strategies 80–81
breeds
  and forage efficiency vi, viii, ix, x
  Ayrshire x
  Canadienne ix
  Dexter vi
  Dutch Belted 80, vii
  Dutch Friesian 80, 82
  Guernsey 60, ix
  Holstein 10–11, 60, 79, 83, 100, vii, viii
  Jersey 35, 60, 80, 121, 130, 155
  milking shorthorns 80, viii
  Red Poll 82, vii
Brucella abortus 62
butterfat 4, 60–61, 71, 83, 100, vii, ix
  and equipment cleaning 66
Butterworks Farm 155

C

calcium
  as mineral nutrient 24, 26–27, 46, 51, iv
  as soil nutrient 87–88
  in composition of milk 60
  ratios in feed 24
calcium carbonate 27–28
calf
  digestive stress 38
  feeding strategies 35
  health 29, 35
  housing 31
  immune system development 34
  nutrition 32
  rumen 34
California Mastitis Test 63
calving 23–26
  and cow health 25, 144
  records 149
Canada thistle 112, 128
carbohydrates
  and nutrition 6, 9
  in plants 12–13, 98
  metabolism iv–v
certification 134–150
  approved materials 142–143, 146

  process 134–137
  records 2, 4–5, 134–137, 144, 146–150
certifying agent 31, 102, 134–141, 143–152
chlorhexidine 143
chlorine 67–68, 143
  as mineral nutrient v
Christianson, Cheyenne 83–84
cobalt v
coliform count 70
colostrum 26, 29, 30–31, 36, 55, 77
conjugated linoleic acid (CLA) 130
conversion to organic dairy farming 4, 144, 148
copper
  as mineral nutrient v
  as soil amendment 140
  as soil nutrient 87
copper sulfate 143
corn silage 10–11, 14, 24
  and crop rotation 91
  overfeeding 24
cover crops 98, 103, 106, 109–110, 112, 115, 117, 140
crop pests
  bacteria 109
  insects 109
crop rotation 86, 89, 96, 99, 102–105, 109, 111–112, 114, 119, 133, 136, 140
crossbreeding 82–83
cryoscope test 72
Crystal Cream and Butter Co. 153
culling 4, 40, 55, 63, 76–77, 101, 149
cultivation 89, 92, 100–101, 108, 110, 114–116, 117, 141

D

dairy processing 59, 134, 138–139, 151
diarrhea 37–38, 55
  in humans 62
  scours 33–39, 51
diphtheria 62
displaced abomasums 6, 24
dry cows 23–26, 121

ration 22
treatments 26
dry matter 9–11, 120–122
   as feed supplement 17
dry periods 15, 23, 26, 27, 124
dysentery 37

E

electrolytes 143
energy 6, 12–13, 15, 17, 47, 120, 123–124, v
   and mineral deficiency iv
   energy-to-protein ratio 16, 19
   in plants 88, 128
   protein-to-energy ratio 20
   requirements for calves 32, 34, 36
energy source 9
energy values 11, 18
Enterobacter 48–49
equipment cleaning 136, 144
   and butterfat 66
   milkhouse 65–69
Escherichia coli 37, 44, 47–49, 62
essential oils 44–45, 47, 144
eucalyptus 44–45, 47
euthanasia 50, 57–58

F

fats
   digestion of 34
   equipment cleaning 65
   metabolism of v
fatty acids 9, 34
feces 37
   bacteria in 49
feed preservatives 13
feed quality 10, 12, 146
feed supplements 4, 51, 136, 142, 146
   ATTRA form xi
fencing 39, 120
fermentation 8–9, 69
   and silage inoculants 13–14
fly control 50
forage 8–11, 46–47, 79, 83, 127
   and calves 34–36
   and crop rotation 99, 101, 104, i–ii
   and weed management 114
   availability 122

quality 12, 14–15, 24–25, 85, 118, 120–121, 125, vii
types of 88–89, 91–92, 117
yield 123
forage-to-grain ratio 16, 19, 22
forage soybeans 112
foundation breeding 80
freezing point 72
freshening 16

G

galactose 60
garlic 26, 43–44, 46–47, 61
genetics 7, 9, 18, 61, 79–83, 101, 144, 147
genetic engineering 115, 135, 142
gestation 15, 142–143, 145, 150
glucose 47, 143, v
   in milk 60
glycerin 44, 143
goldenseal 44, 47
grazing 130–131, 133, 141–142, 149, iii, vii
   rotational 4, 130, 149, vii, ix
Greenberg, Jim 73–75
green manure crops 87, 91–92, 97, 99, 109, 115, 126, 140, i
gypsum 88

H

hay 28, 120–121, 130–131
   for calves 33, 35, 38, 39
   yield iii
hay cutting 133
homeopathy 25, 44–45, 47, 74, 144, 147
Horizon Organic Dairy 3, 75, 153
housing 42, 50, 63, 137, 139, 142–144, 149
hydrogen peroxide 67, 143

I

immune system 7, 24–27, 31, 38, 41, 44–48, 77, 144
insects
   beneficial 115–116, 119, 132–133
   management 86, 89, 102, 105, 107, 109–111, 114–115
interseeding 111
iodine 28, 84, 143, v

as teat dip 67
as topical treatment 29
iron 51, v
   as soil nutrient 87

J

Johne's disease 31, 37, 76–77

K

kelp 26, 28, 46, 84
   as soil amendment 87

L

lactation 11, 15, 22, 25, 39, 51, 81, iv
   feed ration 17
Lazor, Jack 155
legumes 89, 97–98, 103, 115, 117, 124, 126–129, i, iii
   alfalfa ii
   and weed management 127
limestone
   as feed supplement 27–28
   as soil amendment 84, 93, 140
   in feed ration 17
liniments 45, 47
Listeria monocytogenes 62

M

magnesium
   as mineral nutrient 28, v
   as soil nutrient 87–88
   sources 88
manganese
   as mineral nutrient v
   as soil nutrient 87
manure 55, 76, 88, 91, 94, 97–99, 101, 105, 109, 125, 127, 130, i
   application of 100, 126
   application timing 141
   as nitrogen source 125
   as soil amendment 93
   bacteria in 49
   composting of 89–90
   feeding prohibition 142
   handling 131
   in housing 63
   management 7, 86, 137–139, 141, 143
   on hay fields 89
   pack 74, 84

mastitis 59, 150
  and milk quality 63, 69, 72
  causes of 48–49, 62
  prevention of 64, 76, 79, 81–82, 101
  symptoms of 42
  treatment of 26, 40, 45, 47, 148
milk cooperatives 152
milk fat 81
milk fever 26
milk pricing 4, 75, 103, 139, 146, 152–156
milk production 6, 8, 16, 48, 50, 80–81, 120, 124, 127, 130, 152, 154, vii, ix
  costs 149
  regulation 156
milk quality 35, 59, vii
milk replacer 31–32, 35–36, 142
milk tests 69–72
  acid degree value 71
  coliform count 70
  cryoscope test (freezing point) 72
  pasteurized count 70
  preliminary incubation test 70
  sediment 71
  standard plate count 69
minerals
  and equipment cleaning 66–67
  and forage 15, 25
  and milk composition 60
  and nutritional needs 6, 9
  and soils 83, 85–89, 98, 127
  as feed supplements 26–28, 124–125, 146
  fertilizers 97, 100, 140–141
  free choice 24, 46, 84
  management in soil 86–88
mineral cycling 120, 124–125
mineral oil 143, 146

N

National Organic Program 2, 86, 89, 110, 127, 132, 134, 138–139, 149, 152
Natural by Nature 153
nutrition 8–40
  dairy cows 5, 6, 77, 85, 144
  of crops 88, 127
  of forage 80, 121, 123
  of organic food 152

O

on-farm processing 152–155
Organic Choice 153
organic farm plan 135
Organic Foods Production Act 2, 134–135
Organic Materials Review Institute 141
organic matter 85, 87, 109, 125, i–iii
  and soil tests 94–96, 124
  loss 88–89, 91
  management 96–100, 102–103, 131, 141
Organic Valley 3, 62, 100, 153
outdoor access 142
oxytocin 143

P

parasites 77
  prevention of 82
pasteurization 62, 70, 155–156
pasteurized count (LPC) 70
pasture 4–5, 28–29, 39–40, 118–129
  access to 63, 73–75
  and breeds 81, vi–x
  and cow behavior 52
  and milk quality 72
  and parasites 82
  as forage 10, 17, 122
  fertility 124–125
  NOP requirements 6, 127, 136–137, 141–144, 146, 149, xi
Pastureland 154
pasture management 84, 96, 98, 110, 118–133, iii
pasture plate 122
pasture probe 122
pay programs 153–155
peroxyacetic acid 67
Peters, Wayne 100–101
pH
  of properly stored feed 14
  of soil 87–88, 94–95, 124, 126, 129
  of the rumen 8, 46
phosphorus
  and acid sanitizers 67
  as feed supplement 27
  as mineral nutrient 24, iv–v
  as soil nutrient 87–88, 95, 124
  availablilty in soil 98, 105
  functions v
  ratios 16
pneumonia 33–34, 38, 51, 62, 144
potassium 24, 89, v
  as feed supplement 28
  as soil nutrient 87, 95, 124, 140
  to calcium ratio 24
potassium bicarbonate 28
potassium chloride 28
probiotics 26, 44, 46–47, 144, 148
prohibited chemical sanitizers 67
propionic acid 13–14
protein
  and equipment cleaning 65–67
  and forage 15, 25, 121, 127
  and milk composition 60–61, 81, vii, ix
  and milk production 74, vii
  and milk quality 62–63
  as nutrient 6, 9, 11, 16, iv
  excess 49
  ratios 16, 19
  sources for calves 33, 35

Q

quarantine 77
quaternary ammonia 67, 144

R

Redmond Natural Salt 24
respiratory problems 25, 38
  mullein as aid in 45–46
retained placenta 150, v
rock phosphate 87–88, 140
root stress 96
roughage 10, 12, 142
rumen 8, 10, 11, 36, 47
  pH 8
  undeveloped in calves 34
rumination 8–9

S

salt 128
  in feed ration 17
  in soils 95–96
  sources 24, 26, 28, 146
sanitation 26, 50, 69, 81, 143
  field 115, 141
  milkhouse 65–69, 146

scarlet fever 62
sediment 71
silage inoculants 13, 136
smother crop 111
sodium bicarbonate 28
soil management 109, 112
soil nutrition 6
soil quality 86, 97, 115, i–ii
soil tests 84–85, 87–88, 92–96, 105, 124, 129, 136, 140
  grid sampling 93
  laboratories 94
  zone sampling 93
somatic cell 36
somatic cell count 26, 39, 59–60, 62–63, 137, 149
Springfield Creamery 153
St. Johns wort 47
Staphylococcus 62–63
  as mastitis cause 48–49
Stonyfield Farm 153
Streptococcus 62
  as mastitis cause 48–49
stress 5, 7, 16, 22, 24, 27, 30, 40–42, 45, 77, 143, 147
  and cold exposure 57
  and handling 50–58
  and transporting 54–57
  as mastitis factor 48
  from isolation 52–53
  in calves 31, 33, 36, 38
  physiological 51
  psychological 51
  reducing 82

T

teats 49, 64, 70
tillage 86, 98, 104, 109–112, 117
  maintaining equipment 106
tinctures 44, 47, 148
Total Digestible Nutrients 11–12
Total Mixed Ration 10
trace macro elements 44, 46
tuberculosis 62

U

udder 25–26, 30, 42, 47, 61, 64, 70–71, 81, 148, ix
  and milk quality 62
  health 48–49, 63
udder edema 24
udder wash 64, 67
Undulant Fever 62

urine 126–127
uterus 44

V

vaccination 25, 42, 77
ventilation 42, 144
  during transport 56–57
vinegar 67, 128
vitamins 6, 44, 142, 146
  function iv
Vitamin A 28, 51, iv, ix
Vitamin C 46–47
Vitamin E 51, 142, iv
vulva 44

W

water 50
  abnormal amounts in milk 72
  access to 7, 24, 36
  and behavior 53
  and coliform testing 49
  and minerals 51
  and nitrates 42, 49
  and sanitation 64, 65, 66–67, 70–71, 156
  and soil erosion 99–100, 103, 124
  milk component 60
  quality 86, 89, 96, 118, 132–133, 138, 141, 142
waterers 77, 120
  watering fountains 74
weaning 38–39, 51
weed
  management 112–114, 127–128
  suppression 113, 117
whey 26, 44
  in milk 60, 63
Wisconsin Dairy Graziers Cooperative Dairy 154
Wisconsin Organics 154

Z

zinc
  as feed supplement 51
  as mineral v
  as soil nutrient 87